Nocturnal Animals and Classroom Nights

Other Books by Barbara Dondiego

After School Crafts

Crafts for Kids: A Month-by-Month Idea Book, 2nd edition

Year-Round Crafts for Kids

Other Books in this Series by Barbara Dondiego and Rhonda Vansant

Cats, Dogs, and Classroom Pets: Science in Art, Song, and Play

Moths, Butterflies, Other Insects, and Spiders: Science in Art, Song, and Play

Seeds, Flowers, and Trees: Science in Art, Song, and Play

Shells, Whales, and Fish Tails: Science in Art, Song, and Play

Nocturnal Animals and Classroom Nights

Science in Art, Song, and Play

Rhonda Vansant, Ed.D.

Barbara Dondiego, M.Ed.

Illustrations by Claire Kalish

Science in Every Sense

Connecting kids, parents, and teachers through learning

An imprint of McGraw-Hill

New York San Francisco Washington, D.C. Auckland Bogotá Caracas
Lisbon London Madrid Mexico City Milan Montreal New Delhi
San Juan Singapore Sydney Tokyo Toronto

Library of Congress Cataloging-in-Publication Data

Vansant, Rhonda.
Nocturnal animals and classroom nights / Rhonda Vansant, Barbara Dondiego
p. cm.
Includes index.
ISBN 0-07-017911-5
1. Nocturnal animals—Study and teaching (Early childhood)
2. Night—Study and teaching (Early childhood) 3. Nocturnal animals—Study and teaching—Activity programs. 4. Night—Study and teaching—Activity programs. I. Dondiego, Barbara L. II. Title.
QL755.5.V36 1997
372.3'5044—dc21 97–5406
CIP

McGraw-Hill

A Division of The **McGraw-Hill** *Companies*

1 2 3 4 5 6 7 8 9 0 BBC/BBC 9 0 2 1 0 9 8 7

ISBN 0-07 017911-5

The sponsoring editor for this book was Judith Terrill-Breuer, the editing supervisor was Sally Glover, and the production supervisor was Suzanne Rapcavage. It was set in Giovanni Book by Jaclyn J. Boone of McGraw-Hill's Professional Book Group in Hightstown, New Jersey.

Printed and bound by Braceland.

SIES/97

Dedication

To Brian Grunwald, a recipient of the Professional Enrichment Endowment for Teaching, which I gave him in memory of my mother and father. He is now my colleague and spiritual brother, who understands my sincere desire to help children love and care for this world, and to whom I am known as "Joy."

With love,
R. Joy

To Mary, who adds so much brightness to all my days.

With love,
Mom

Contents

A Letter to Teachers

Dear Teachers,

Maybe you're one of many educators who feels a certain anxiety about teaching science. Perhaps your science courses in school and college were stressful as you plodded through specific experiments, tried to memorize the periodic table, and tried to understand phenomena without the opportunity to build concepts first. You probably have forgotten much of the science instruction because it went only into short-term memory and had no direct link to what you were encountering in your world at that time. Perhaps you never had a role model who felt a passion about loving and caring for the world—a role model who would dare to teach you "how" to learn rather than "what" to learn.

Whether you place yourself in this category or were blessed with good role models who instilled in you a zeal for learning science, we hope this book will contribute to meaningful and enjoyable science instruction in your class.

Because we embrace the definition that science for young children is studying and exploring our world, we strongly feel that science should be the focus of an early childhood curriculum. Although this is a science book, it is also a book about life and learning. It is designed to serve as a framework for a thematic study. Within this framework, you can add your own literature, math experiences, writing experiences, and other activities within content areas. The format of this book allows your teaching and learning experiences to flow together in a natural, integrated way.

We hope that you find this format comfortable for teaching science and that you enjoy, with your students, the journey of discovery. Perhaps you will be one who inspires this generation of children to find joy and excitement in learning about the wonders that surround us.

Sincerely,

Rhonda Vansant

Barbara L. Dowling

A Letter to Parents

Dear Parents,

Generations of children have grown up feeling that science was too difficult, too stressful. Many children have avoided science and ranked it among their least favorite subjects.

How wonderful to think that we might change that attitude for the current generation! We, the authors, view science as studying and exploring our world in ways that are appropriate for the learner's stage of development. Children explore the world quite naturally, and we want to build on the natural inclinations of children by guiding their explorations and nurturing their curiosity. When children see their ice cream melt, they have an opportunity to learn about their world. When children feel the wind blow, they are discovering information about their world. We do not have to search for expensive equipment to teach science to young children; we simply need to take advantage of everything that is already around us.

One of the most precious gifts we can give our children is a love of learning. We hope you will find ways to use this book with your own children as well as with groups of children in various organizations. It is our hope that you may nurture your children's natural wonder and curiosity about our world and that the dream that this generation will come to love, understand, and cherish this world will come true.

Sincerely,

Rhonda Vansant

Introduction

Nighttime and the study of nocturnal animals are appropriate topics for young children because they are curious about the dark and what happens while they are asleep. Sometimes children have fears about night because they don't know the many wonderful and natural things that occur then. This book can help children learn about night in an enjoyable and meaningful way. They can come to respect the animals that eat and work at night and appreciate how wonderfully our world functions, both day and night.

Building concepts

Concepts are built through *concrete experiences* with real objects or living things. Concepts are the foundations for subsequent learning. After children experience reality, they can then make models or re-create the experience in a variety of ways. Models and pictures are *semi-concrete representations*. Words that we attach to these experiences are *symbolic representations*. The word *sun* written in a book symbolizes an object that we see almost every day. Giving children opportunities to experience real living and nonliving things helps them develop concepts that, in turn, give meaning to the written and spoken word.

Nocturnal Animals and Classroom Nights provides children with opportunities to build concepts as they observe the sun, moon, stars, and nighttime in a scientific way. Children may not get to see real nocturnal animals, so it is very important that the teacher provide many informational videotapes and books and that the children make the models of these animals.

The learning activities in this book include the following:

- ☐ **Art**—Creating lifelike models
- ☐ **Creative drama**—Acting, pretending
- ☐ **Music and dancing**—Singing songs and moving creatively
- ☐ **Informational books**—Finding facts
- ☐ **Research**—Using the five senses to discover information; reading and looking at pictures; talking to people
- ☐ **Writing**—Preparing factual reports or creating essays, poems, or stories to express thoughts and emotions
- ☐ **Cooking**—Using a variety of skills to create treats
- ☐ **Mathematics**—Measuring, counting

Each chapter in this book presents factual information you can read aloud to the children. You might want the children to sit on the floor near you while you share this information. If you keep a chart of important words for the children's future writing, place it nearby so that you can add words throughout your reading and discussion time.

Teaching children science

For young children, science should not be a set of experiments with specific steps that must be followed. It should involve a very natural discovery of their world through real experiences, creative art, literature, drama, music, writing, reading, and play. Whenever we teach children to use their five senses, we are teaching science. Whenever we provide opportunities for exploration and discovery, we are teaching science. Whenever we help children get to know the world around them, we are teaching science. Whenever we teach children to love and care for the world, we are teaching science.

The following science skills are appropriate for instruction with young children. You will be helping children use these skills as you pursue your study.

1. *Observing*—Using any of the five senses to become aware of objects
2. *Following directions*—Listening to or reading step-by-step directions and carrying them out
3. *Classifying*—Arranging objects or information in groups according to some method
4. *Creating models*—Portraying information through multisensory representations
5. *Manipulating materials*—Handling materials safely and effectively
6. *Measuring*—Making quantitative observations (time, temperature, weight, length, etc.)
7. *Using numbers*—Applying mathematical rules
8. *Asking questions*—Verbally demonstrating curiosity
9. *Finding information*—Locating words, pictures, or numbers
10. *Making predictions*—Suggesting what may happen (Predictions should come after children have some experiences with the topic. Predictions should be based on previously gathered data.)
11. *Designing investigations*—Coming up with a plan to find out information or answers to questions
12. *Communicating or recording information*—Communicating or recording information by
    - Talking to the teacher and/or other children
    - Playing with theme-related props
    - Drawing pictures
    - Labeling
    - Making diagrams
    - Making graphs
    - Writing (descriptions in learning logs or narratives)
    - Taking photographs
    - Recording on audio or videotape

13. *Drawing conclusions*—Coming to various conclusions based on their stage of cognitive growth and their prior experiences
14. *Applying knowledge*—Finding ways to use what is learned

About this book

Each chapter starts with a science goal to guide parents and teachers. The goal is followed by ideas on how to plan for the activities, as well as lists of visual aids and related words that can be used for discussion.

Safety symbols, or icons, have been placed in the text to alert parents and teachers about activities that require supervision or other precautions.

 Scissors

 Adult supervision

Discussion Ideas, Activity Ideas, and Science Skills are also indicated by symbols in the text:

 Discussion Ideas include questions for discussion that the teacher can ask the children.

 Activity Ideas incorporate multicurricular activities, such as art, music, and drama, that can be easily planned ahead of time and implemented throughout the study.

 Science Skills detail exactly what that child will learn by performing the activity.

Nighttime Notes and Let's Create are interspersed throughout the text. Nighttime Notes contain interesting tidbits of information to be passed along verbally to the children or instructional information for teachers or parents. Let's Create ideas include directions and simple patterns so the children can build actual models of the animals being studied. Children personify objects quite naturally. They enjoy giving their models names and personalities as they use them in a variety of ways. All of the Let's Create ideas can be embellished with a child's creative thoughts. The models can then be used in preplanned activities as well as spontaneous play.

A final word

For parents and teachers, every moment is an opportunity to teach science. By using this book, you can help children better understand the wonder of night. We hope you enjoy using this book and that your children's sensitivity to our world is enhanced.

Nocturnal Animals and Classroom Nights

chapter 1

Night and Day

Science goals

To help children better understand day and night and their relationship to each other and to make them familiar with the sun, moon, and stars

Planning

Gather informational books, posters, videotapes, and computer software from the library or media center. Place the books in an area where they are readily available to the children. Have chart paper available. Some homework sheets are included in this chapter. Make copies of them if you want to use them.

It is suggested later in this chapter that you invite people who work at night to come in to talk to the children about their nighttime jobs. You may want to send letters to possible guests to arrange their visits.

Materials needed for discussion and activities

- ☐ Chart paper
- ☐ Clipboards and paper for writing
- ☐ Quilt
- ☐ Copies of letters to send home
- ☐ Large and small balls
- ☐ Globe
- ☐ Small moon made of clay
- ☐ Construction paper
- ☐ Chalk
- ☐ Paper punch
- ☐ Shoebox
- ☐ Flashlights

Related words

day The 24-hour period during which the earth rotates once on its axis is one definition, but for this book we use another accepted definition, which is the period of time in which there is light (between dawn and the end of sunset).

night The period of darkness between the end of sunset and the beginning of dawn.

stars Celestial bodies that look like twinkling points of light when seen at night from earth.

This book can help us learn about nighttime. Most of us are asleep when it is night, so we do not realize all of the marvelous things that are happening when darkness comes. In this chapter we will learn about how and why we have both day and night. We will learn about the earth, sun, other stars, and the moon.

Each day of the week is 24 hours long. Some of these hours are called "day" and some of these hours are called "night."

Day and night

Classifying; asking questions

Help children explore their own ideas about day and night. Make two columns on chart paper. Write "Day" at the top of one column and "Night" at the top of the other column. Let the children think of words that they associate with each, and you can write them under the headings. Examples might be:

Day	**Night**
sun	stars
play outside	moon
swim	dark
eat breakfast	sleep
eat lunch	bed
go to school	dreams

Another way to explore day and night is to let the children tell you activities they do during the day, night, or both. Write these on the chalkboard or on chart paper.

Day	**Night**	**Both**
eat breakfast	go to sleep in my bed	read books
eat lunch	watch the stars	eat
go to school		

Now ask the children to tell you some things or ask some questions to let you know what they would like to find out. Record these questions on a chart so that you can refer to them as you move through this unit of study.

Take a daytime walk outside

Observing; recording information

Plan a time to take your children outside so that you can talk about the earth, the moon, the sun, and other stars. Take paper and a clipboard so that you can write about the experience as it proceeds. Take notes on what the children see and what they say. As you walk on the ground, tell the children that they are walking on the earth. The earth is a giant ball that is covered with soil, rocks, plants, and water. The earth is surrounded by air. Let the children tell you some things they are observing about the earth.

Now ask the children to look up into the sky, and ask them what they see. You may see the moon if it is early morning. You may see a blue sky or a gray sky. You may see the sun. Remind children that they should never look directly into the sun. It is very harmful to the eyes. They should only glance quickly toward the sun. Let the children talk about the sun. Stand in a designated place and help the children decide where the sun seems to be and record this information on your paper. For example, you may say that the sun is behind the big oak tree on the playground.

You may see clouds. Let the children talk about the clouds. Ask the children if they see stars (other than the sun).

Ask the children to lie down on their backs on the grass. You may want to provide a quilt or towels to lie on. Ask the children to lie very still and quietly and look all around them. Then ask, "Is anything moving?" (Clouds may be moving. The wind may be blowing leaves. Encourage the children to discuss any movement they see.)

Before you go inside, go back to the designated place where you observed the sun before. Check on the sun now. Is it still in the same place? Record this information on your paper.

Assignment for home

Observing; recording information

In preparing to help children learn about night, it is important to activate prior knowledge and incorporate their experiences.

Prepare an assignment to be done by each child and a parent at home. The activity is called "A Nighttime Observation." Duplicate the assignment sheet on the next page and send it home on a convenient night. Have a designated time for it to be returned.

Up close and far away

Observing; following directions; designing investigations; drawing conclusions

In studying objects in space, it is very important that children understand that objects far away look smaller and objects that are closer look larger. Do one or more of these activities with your children. Then guide them to conclude that objects look smaller when they are far away from us.

1. Get a large round ball, like a basketball. Let one child hold it out in front of the children (the child could stand about 2 feet from the other children). Ask the children to hold their hands apart to show how big the ball is. Then let the child take the ball about 30 feet away. Let the children use their hands to measure again. Let the child take the ball as far away as is safe and possible. Let the children measure again. (They will probably have to use their fingers this time because the ball will look so tiny.)
2. Repeat this activity with a tennis ball.
3. Repeat this activity with just the person.
4. Let children look down the street and talk about how houses and buildings look smaller the farther they are away from the children.
5. Older children could draw pictures in which they show objects that are close up to be larger and objects that are far away to be smaller.

Flying in an airplane

Applying knowledge

Ask if any children have ever flown in an airplane. Ask questions like "How did houses and cars look? Could you see people walking around?" and "Why do you think everything looked so small?"

Night Notes for Home

Dear Parents,

We are doing a unit of study about nighttime. To build important concepts, each child needs to go out one night this week to observe night. Please go outside with your child to complete this assignment, and you can write the answers or help your child write and draw. Thank you.

Please return this form to school on ______________________.

Child's name ______________________

Date ________________ Time ____________________

Look up into the sky. What do you see?

____

____

____

Write "yes" or "no" to answer the following questions.

Do you see clouds? ______

Do you see stars? ______

Do you see the sun? ______

Do you see the moon? ______

If you see the moon, draw a picture of it here.

Do you see anything else in the sky? ______

If you do, what do you think it is? ______________________________

Look around you as you stand here on earth. How do things look?

____

Be very still and quiet. What sounds do you hear?

____

Our world is a giant ball. It is one of nine planets that travel through space around the sun. The sun and everything that travels around it make up the *solar system*. The moon or moons that move around the planets are included in the solar system (Fig. 1-1).

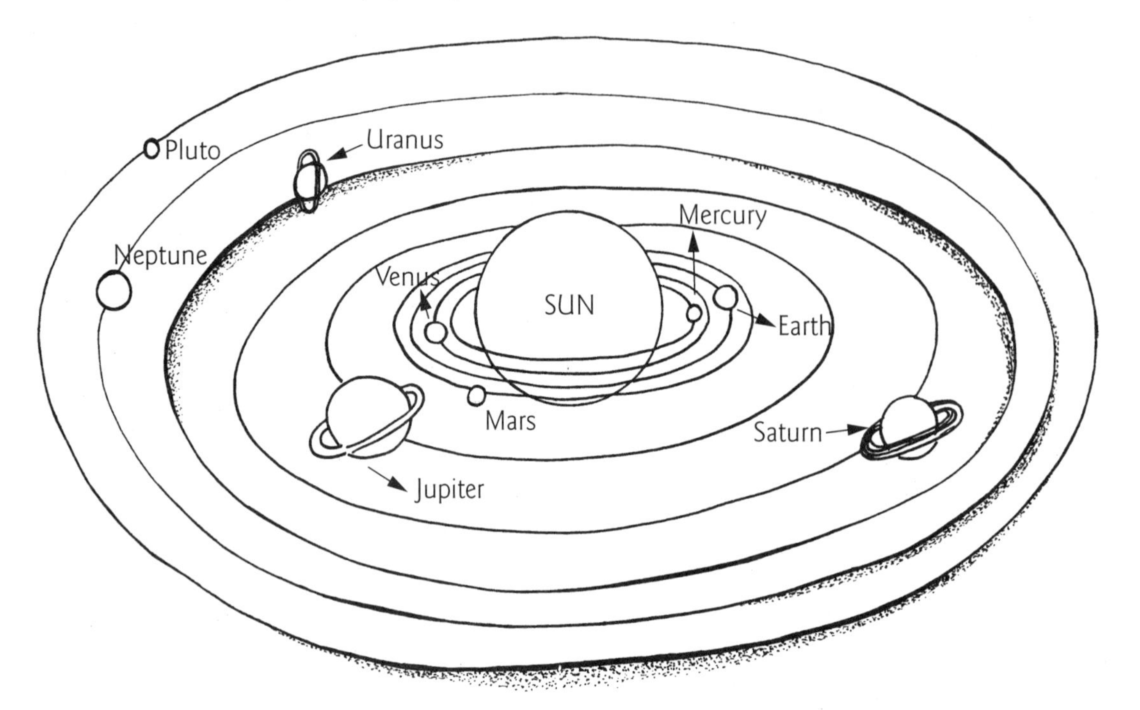

■ **1-1** *The sun and planets.*

The nine planets, in order of increasing distance from the sun, are Mercury, Venus, Earth, Mars, Jupiter, Saturn, Uranus, Neptune, and Pluto. All of the planets move around the sun in paths called *orbits*. Some of the planets (Venus, Mars, Jupiter, and Saturn) can be seen without a telescope. They look like bright stars in the sky.

Nighttime Note *Depending on the age of your children, you may or may not want to discuss the planets. The information about the solar system is more for the benefit of the teacher. Much of this information is too abstract for young learners, but it can help you, the teacher, as you guide the children in their learning.*

☞ *Day and night*

Creating models; designing investigations

Provide a globe that the children can look at and hold. Also provide a lamp with the lightbulb exposed without a lamp shade. This light serves as the sun in our model. Use white modeling clay to make a round moon (about the size of a tennis ball).

Help the children look at the globe first. Show them the land and water. Show them the North Pole and the South Pole. Show them the equator and any other aspects of the globe that you feel are appropriate for your children. Show them where they live. Tape a colored piece of construction paper on that spot.

It is hard to believe, but the earth is always moving. It is spinning around like a top. Use the globe to show the children that the earth is turning (toward the right, or eastward) all the time. It is also traveling around the sun. This path is called the *earth's orbit*. Demonstrate this orbit by walking around the light with the globe. The movement of the earth helps us determine days and years. It takes one day, or 24 hours, for the earth to spin around one time. It is daytime where we are when the sun is shining on our part of the earth. It is nighttime when the sun is not shining on our part of the earth. (Help the children see this with the help of the globe and the light.)

It takes one year (365 days) for the earth to travel around the sun one time. Because the earth is tilted and because of its path around the sun, we have seasons—winter, spring, summer, and fall.

The sun is a huge, glowing ball of gases. The sun gives us heat and light. The sun is a star. All stars are made of glowing gases. It is the closest star to us, so it looks much bigger than other stars, but it is not the biggest star; in fact, it is a medium-size star. It is in the middle of the solar system, and all of the planets travel around it. The diameter of the sun is more than 100 times the diameter of the earth. The sun is about 93 million miles from us.

It seems that some days the sun is very bright, and some days it is not bright, and some days we cannot even see it. Weather conditions cause these differences to happen.

Nighttime Note *The sun is one of billions of stars that make up a galaxy. The particular galaxy we live in is called the Milky Way. The Milky Way turns also, and the solar system travels around the center of the Milky Way. It takes 225 million years to make one revolution around the center of the Milky Way. The sun spins just like the earth spins, but it takes the sun much longer to spin around once.*

The moon is an object that travels around the earth. The moon is made of dust and rock. It has tall mountains and flat areas. Big holes on the moon are called craters.

The moon seems to be larger than the stars, but it seems larger because it is so close to us. The diameter of the moon is about one-quarter the diameter of the earth. It takes the moon $29\frac{1}{2}$ days to make its trip around the earth. The moon rotates completely around once during each trip around the earth. The moon travels west to east, just like the earth's rotation. The moon looks very bright in the sky at night, but it actually has no light of its own. The moon reflects light from the sun. (Use a mirror and the light to show children how light can be reflected.)

Decide on a way to use your globe, light, and clay moon to demonstrate day and night. Since you labeled the area where you live on the globe, the children will be able to see how they are having day while the other part of the

world is having night and how they are having night while the other part of the world is having day. This concept is difficult to develop for young children because we have to rely on small representations, but at least the children will have an idea about how and why we have day and night.

Lunar phases

Creating models

Because the moon is moving around the earth, we see a different part of the moon each night. It seems like the moon is changing its shape, but it really isn't (Fig. 1-2).

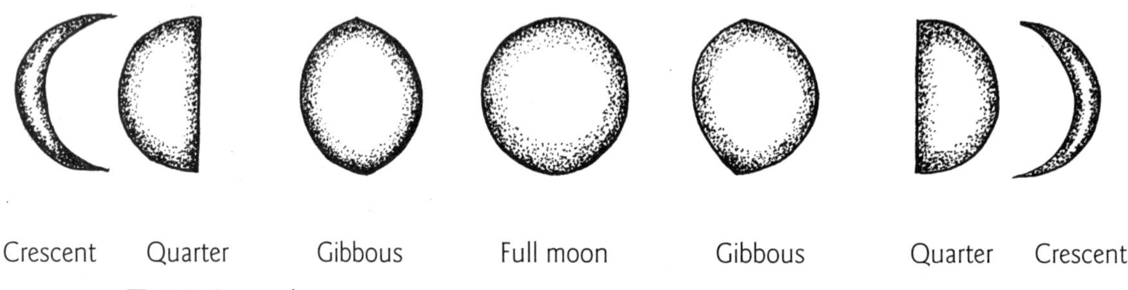

■ **1-2** *Lunar phases.*

Use the clay moon you made to help children understand better how the moon moves. Hold the moon so that the light from the lightbulb is shining on it. Now let the children walk around the moon so they can see it from different angles. Discuss what they see.

A moon diary

Observing; following directions; recording information

Prepare a small booklet for moon observations for each child. Each booklet should be about 4 inches by 5 inches. If you want the children to observe the moon for a month, the booklets should have 30 or 31 blank pages (or however many days there are in the month you want to do the observations). If you want the observations to last for a week or two weeks, adjust the number of pages so that there is one page for each day of observation. The cover of the booklet should say "My Moon Diary." Each page should have a place for the date. Tell the children that each night for the designated period they need to look at the moon and draw a picture of what it looks like.

If they cannot see the moon at all, they should leave the page blank. Send a note home to encourage parents to assist with this project. At the end of the observation time, let the children bring their booklets back, and you can discuss their findings.

Moon pictures

Applying knowledge

After the children have seen the various shapes of the moon, let them create night pictures that include the moon. Provide black or dark blue

construction paper and white and colored chalk. They can use the white chalk to draw stars and moon and the colored chalk to add houses or people or cars or trees or whatever they would like. Encourage the children to use different phases of the moon.

Another idea for interesting pictures is to provide a piece of white construction paper and crayons for each child. The children should draw and color a night picture (they should color the moon with a white crayon). After the picture is colored, they can paint over the entire picture with a thin blue or black paint. The crayon areas will resist the paint, and the rest of the paper will absorb it so a nighttime effect is created.

Stars are balls of glowing gases (mainly hydrogen and helium). Stars are in the sky all of the time, but we can see them better at night. During the day, the light of the sun makes it hard for us to see the other stars. Stars seem to twinkle at night. This twinkling occurs because the starlight moves through moving air that surrounds the earth. Some stars look brighter than others. Stars have different colors. Some look yellow, some look bluish, and some look reddish.

Stars seem to move, like the sun seems to move, but the reason they seem to move is because the earth is spinning. Stars are very far away. Stars are millions and millions and millions of miles away.

> **Nighttime Note** *Stars are so far away from the earth that the distances are measured in light-years. A light-year is the distance light travels in a year. Light travels 186,000 miles per second and 5,880,000,000,000 miles in a year.*

☞ ***Looking at stars***

Observing; finding information; recording information

People have always been interested in studying the stars. People who lived long ago made up stories about people or animals or objects they thought the stars looked like.

☞ ***Star pictures***

Creating models

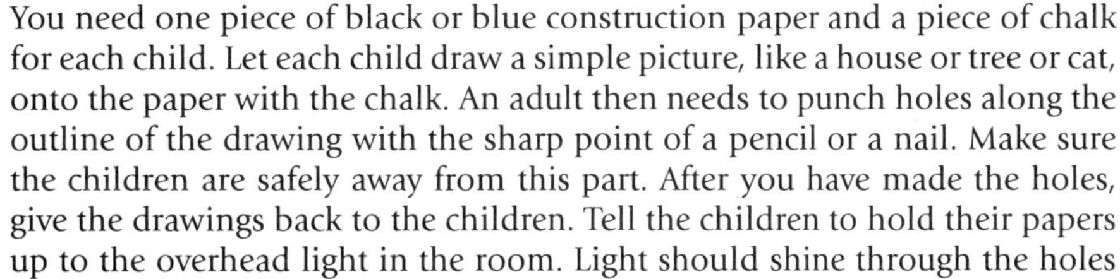

You need one piece of black or blue construction paper and a piece of chalk for each child. Let each child draw a simple picture, like a house or tree or cat, onto the paper with the chalk. An adult then needs to punch holes along the outline of the drawing with the sharp point of a pencil or a nail. Make sure the children are safely away from this part. After you have made the holes, give the drawings back to the children. Tell the children to hold their papers up to the overhead light in the room. Light should shine through the holes and look like stars in a constellation.

Another way to create star pictures is to punch holes in the bottom of a shoe box. Turn the lights off in the room. Turn the shoe box over and hold a flashlight so the light shines inside the box and through the holes.

Night Notes for Home

Day of the week ____________________ Time ____________________

Dear Parents,

Please go outside one night this week with your child and help him or her look at the stars. Please complete the following as your child talks to you.

Tell me about the stars you see. ____

____

____

Do the stars seem to twinkle? _____________

Are some stars brighter than others? _____________

Point to one that looks brighter than the others.

Look all around the sky. Which part of the sky do you like the best?

____

Why do you like that part? ____

Say or sing "Twinkle, Twinkle, Little Star" together:

Twinkle, twinkle, little star,

How I wonder what you are.

Up above the world so high,

Like a diamond in the sky.

Twinkle, twinkle, little star,

How I wonder what you are.

The person who wrote this poem thought a star looked like a diamond. What do you think a star looks like?

____

____

____

____

Let's create glitter-glue constellation pictures

Following directions; creating a model; manipulating materials; finding information

Pretend a blue or black sheet of construction paper is the night sky. You can create bright constellations with your own glitter-glue (Fig. 1-3).

What you need

- ☐ Waxed paper
- ☐ White glue
- ☐ Gold or silver glitter
- ☐ Round toothpicks
- ☐ Dark blue or black construction paper
- ☐ White chalk

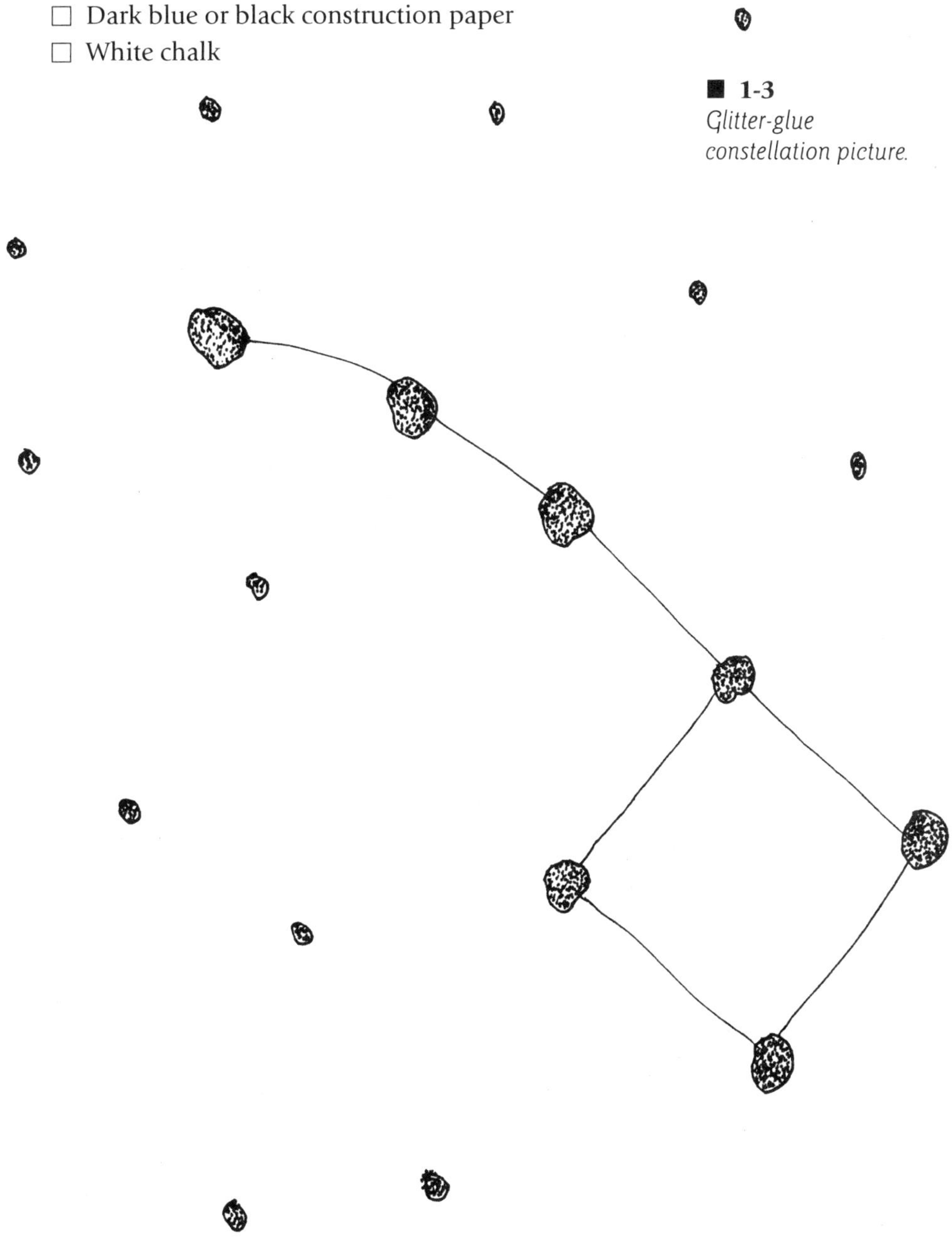

■ **1-3** *Glitter-glue constellation picture.*

Directions

1. Decide which constellation you would like to create. Mark the position of each major star in the constellation with a little chalk dot on your dark construction paper.
2. Make glitter-glue by squirting a small amount of white glue onto waxed paper and adding glitter to it. Mix it with a toothpick.
3. Use the toothpick to put one drop of glitter-glue on each chalk dot of your constellation. You can add a few extra "stars" to your picture with smaller glue dots if you wish.
4. Let the picture dry flat for at least two days. Connect the stars of the constellation with chalk lines. Name your constellation.

Extension activities for students ages 6 and up

- ☐ After creating the glitter-glue constellation pictures, use reference materials to find out about other stars, planets, and constellations.
- ☐ Invent some new constellations. You might enjoy writing, illustrating, and acting out an imaginary story about a pretend constellation.
- ☐ Invent a new story for a real constellation.

Let's create a little book of star stories

Following directions; creating a model; manipulating materials; using a ruler; finding information

For hundreds of years people have made up stories about the groups of stars (constellations) that seem brightest in the night sky. Begin this project by finding a book or two about the stars in your media center or library. Can you find a story that might be your favorite?
Now create your own little book of star stories.

What you will need

- ☐ White poster paper
- ☐ Ruler
- ☐ Scissors
- ☐ Dark blue, black, and white construction paper
- ☐ Paper punch
- ☐ Yarn or paper fasteners
- ☐ Small sticker stars or colorful paper punches
- ☐ White glue (optional)
- ☐ White crayon
- ☐ Pencil

Directions

1. Your book should have a poster paper cover, a blue page for each constellation picture, and a white page for each star story. It can be fastened together with yarn or metal paper fasteners.
2. Measure and cut out two squares of poster paper 5 1/2 by 5 1/2 inches for the front and back covers (Fig. 1-4).
3. Create pictures and write stories about the stars and constellations, and place these pages between the two poster paper covers. You can also include your glitter-glue constellation pictures.
4. Punch holes through all the pages of the book, including the covers.

■ **1-4**
The pattern for the little book of star stories.

My

Star

Stories

5. Put yarn or metal fasteners through the holes to hold the pages together (Fig. 1-5).

■ **1-5** *A little book of star stories.*

☞ *Closure*

Applying knowledge; asking questions; communicating information

Bringing closure to your study is important. Review the initial goals, and give the children opportunities to talk about what they have learned. Help them think about how what they have learned has helped them. The following activities can be done at the end of this chapter or at any point you feel appropriate.

1. Make a class graph to show the times the children go to bed at night.
2. Let children bring one of their favorite bedtime books. You can read these aloud or the children can share them with each other.
3. Let children bring something they like to sleep with (like a teddy bear or blanket). If you have rest time, let the children use this item while they rest.
4. Let the children make nighttime drawings. They should draw and color a nighttime scene on white paper, and then paint over the entire piece of paper with a wash made with black paint thinned with water.
5. Invite people who work at night to come visit the class and talk to the children about their work and how they adjust to working at night. Visitors might be people who are:
   - Nurses
   - Doctors
   - Bakers
   - Firefighters
   - Police officers
   - Cab drivers
   - People who clean office buildings at night
   - Truck drivers
   - People who work in restaurants
6. Let children make houses, hospitals, bakeries, and other buildings out of construction paper and paste on yellow windows (as if lights are on). The children can draw pictures on the windows to show what is happening in the house or whatever building the child has chosen. Glue these buildings onto a large piece of dark blue or black bulletin board paper. Let the children add the moon, stars, cars with lights on, streetlights, and people who might be out at night.

chapter 2

Shadows

Science goals

To help children define what a shadow is and to explore shadows in our world, particularly shadows at night

Planning

Gather informational and story books about shadows and locate them in an area of the room that is easily accessible to the children. You need flashlights, so you may want to send a note home requesting that each child bring a flashlight. A homework activity is included in this chapter, so you may want to make copies to send home.

Materials needed for discussion and activities

- ☐ Lamp
- ☐ Globe
- ☐ Overhead projector
- ☐ Flashlights
- ☐ Copy paper
- ☐ Small plastic objects
- ☐ White sheet

Related word

shadow A dark image created by an object that blocks light

In this chapter, we will learn about shadows. Anytime there is light, there can be shadows. A shadow is formed when something that does not let light through comes in front of the light. The shadow is a dark image. At night we see shadows outside caused by street lights, porch lights, and car lights. Inside at night, we see shadows caused by lights turned on in our house.

The shape of the moon is determined by shadows. Half of the moon is always in sunlight, and the other part is always in shadow. Because the earth and moon are moving in their orbits, various parts of the moon can be seen at different times.

Night is a shadow. The sun is shining on one side of the earth, and the other side is in shadow. (Refer back to Chapter 1 for a full explanation of night and day.) Use a lamp without the shade and a globe to demonstrate how part of the earth is in sunlight and the other part is in shadow.

If the sun is shining, we can see shadows outside. If we are in our houses at night with lights on, we can see shadows. If we are outside at night and there are lights on, we can see shadows.

In helping children learn about shadows, we want to use a method of teaching science called *guided discovery*. In guided discovery, we create situations in which the children can "discover" the information, which may not be new to the world but is new to them. Discovering information is highly motivating to learners.

What are shadows?

Asking questions; observing; manipulating materials; designing investigations; drawing conclusions

You need a flashlight with a very bright light or something like an overhead projector that can shine a bright light against a wall. Turn off the overhead light to simulate nighttime. Have the light positioned about 6 feet from the wall. Walk in front of the light. Ask the children what happened. Put a book in front of the light. Ask the children what happened. Turn the flashlight or projector light off. Walk in front of it. What happens? Put the book in front of it. What happens? Let the children talk to you about their own ideas of what happened. Guide the discussion so that you talk about the light not being able to move through you or the book and that is why a shadow was formed. Without the light, those shadows do not form.

Turn the designated light back on. Let each child have a turn to walk between the light and the wall so that they can experience their own shadow as well as the shadows of their friends.

You may want to try holding various materials, like a paper towel or waxed paper, in front of the light source. Help the children explore what happens when some light can come through the object.

Look for shadows outside during the day

Observing; applying knowledge; communicating information; drawing conclusions

Tell the children that now all of you will go outside and see if you can find shadows. If the sun is shining, the children will immediately see shadows all around. Take a walk and talk about the shadows. Look for shadows of the swings, flowers, trees (the shade of a tree is actually a shadow), and the children themselves. "Do the clouds have shadows? Does the school building have a shadow? Do cars have shadows?"

Ask the children to tell you what is causing the shadows. Their responses should include the idea that people and plants and objects are between the sun and the ground and so shadows are formed.

Tell the children to hop. "Does your shadow hop, too?" Tell the children to run. "Does your shadow run, too?" Tell the children to turn around. "Does your shadow turn, too?"

Play a game of Shadow Tag. One child is designated to be "It." The children run around, and the person who is "It" tries to step on their shadows. If a child's shadow is stepped on, that person must sit or squat down. When three people are down, "It" chooses one of the three to replace him or her, and the game continues.

You also should go outside on a day when the sun is not shining. The children won't be able to find shadows or play Shadow Tag. Ask the children to tell you why there are no shadows outside when the sun is not shining.

☞ *Assignment for home*

Observing; classifying; measuring; following directions; finding information; recording information; applying knowledge

Prepare a sheet like the one on the next page for each child to take home. The children and their parents should observe shadows during the day and night.

☞ *Flashlight fun*

Observing; following directions; manipulating materials; communicating information

Send a note home asking parents to send a flashlight to school with their child.

On the day the children have their flashlights, let the children lie down on their backs on the floor with their flashlights. Turn the lights in the room off to simulate night and let the children turn on their flashlights. Let them shine light all around the room. Encourage them to quietly observe what happens. Let five children at a time get up and walk around the room with their flashlights. Let them talk about the shadows they see.

☞ *How can we make shadows longer or shorter?*

Observing; following directions; manipulating materials; communicating information; making predictions; designing investigations

Each child needs a white piece of paper (about 8 by 11 inches), a small object that will stand on the paper (like a plastic bear or dinosaur that is about 1 or 2 inches tall), and a flashlight. Turn off the overhead lights, and have the children turn on their flashlights.

First, instruct the children to hold their flashlight directly over the object. (You should demonstrate.) "How does the shadow look?" (It will look very small.) Instruct the children to slowly move their flashlights to the side of the object and hold the flashlight on the side of the object. "How does the shadow look?" (The shadow should look longer.) Now move the flashlight slowly away from the object while still holding the flashlight on the side of the object. "As you move your flashlight farther from the object, what happens?" (The shadow gets longer and larger.)

Ask the children what they can do to make the shadow so long that it goes off the white paper. Let children experiment with different objects and different sizes of paper. Encourage them to design their own investigations.

Nighttime is a wonderful time to have shadow plays. In your classroom, you can simulate night by turning off the overhead lights.

You need a very bright light, like an overhead projector or bright lamp or flashlight. You need a large, white sheet and something to hang the sheet on or two children to hold the sheet during the shadow play. Position the light so that it is behind the sheet and there is plenty of space for children to walk between the light and the sheet. The audience should be sitting in front of the sheet.

Night Notes for Home

Dear Parents,

We are doing a unit of study about nighttime. We are learning about shadows. A shadow is produced when light cannot pass through an object. A shadow is black or dark gray.

Please help your child look at shadows during the day and at night. Record information below.

Daytime

Date ____________ Time ____________

Is the sun shining? _________

What shadows do you see? ____

____

____

____

Nighttime

Date ____________ Time ____________

Take a walk around your house with lamps and overhead lights on.

What shadows do you see? ____

____

____

____

If the weather makes it permissible, go outside to look for shadows.

What shadows do you see? ____

____

____

____

To help children get used to shadow play, first play a game called "Guess My Object." Each child chooses an object. One at a time, a child goes behind the sheet and holds up his or her object. The audience tries to guess what the object is by looking at the shadow. This activity is fun for the children.

After the children have had fun experimenting with this arrangement, select a story that they can act out. A story like "Little Red Riding Hood" is a good one. Read the story to the children several times. Decide who wants to be each character. Decide what each character will wear. Now read or tell the story and have the characters act it out, without talking, behind the sheet. The audience and characters will love this. Repeat the performance until everyone has had a turn to be in the play. You can also plan productions using puppets or toy animals.

chapter 3

Nocturnal Animals

Science goal

To help children become familiar with animals that find food and are active during the night

Planning

Gather informational books, story books, videotapes, and computer information from the library or media center. Encyclopedias can be very helpful. You may want to look for pictures of the animals discussed in this chapter. Put all of the materials in a convenient place for children to use.

Materials needed for discussion and activities

- ☐ Lamp
- ☐ Calming music
- ☐ Flashlights
- ☐ Red and green cellophane
- ☐ Sticks

Related words

mammal An animal that has a backbone, has hair or fur, is warm-blooded, and whose young are born alive and drink milk from their mother

nocturnal Active at night

After you go to bed each night, many animals are just waking up and are ready to hunt for food or take care of their babies. We usually do not get to see these animals. They are called *nocturnal* because they are active at night. They like the darkness because they do not have to be concerned about the heat of the day, they are safer from their enemies, and it is easier for them to find food when there aren't so many people and other animals around. Nocturnal animals may have huge eyes that can collect a lot of light, or they may have powerful hearing.

Some animals that are nocturnal are raccoons, foxes, skunks, opossums, moths, fireflies, toads, owls, bats, beavers, coyotes, leopards, hippopotamuses, ghost crabs, and sea turtles.

Raccoon

A raccoon is a furry animal that has rings of black on its tail (five to seven rings) and black around its eyes. A raccoon may be grayish in color with some brown or yellow.

The two main types of raccoons are the northern raccoon and the crab-eating raccoon. Raccoons are mammals. Mammals have hair or fur, are warm-blooded, feed milk to their babies, and breathe with lungs. People are mammals too.

Raccoons have five fingers on their paws. These fingers have sharp claws. Their fingers help them climb, swim, dig, and handle food very skillfully. Raccoons are good at opening trash cans or cabinets or getting seeds from a

bird feeder or tearing open a bag of dog food. Raccoons walk like bears, with all four feet on the ground. They can stand up easily on their back feet when they want to reach something.

How do we know that raccoons have visited us?

Asking questions; drawing conclusions

Start the discussion by asking, "Since raccoons are active at night, how could we know that they have been visiting us?" Guide the children in discussing such "evidence" as a trash can lid lying on the ground or a torn bag of dog food (and some of the food missing).

"If there is snow on the ground, could you find tracks?" Raccoons may live in trees or hollow logs. They may make their home in an old barn that no one is using. If an area does not have trees, raccoons may make their home in tall grass.

Raccoons like to eat nuts, seeds, berries, insects, eggs from birds' nests, corn, fruit, and fish. Raccoons like to catch food from streams, and many times they put their food into water before they eat it.

Northern raccoon mothers have babies about once a year. A mother can have one baby or two babies or as many as seven babies. The mother takes good care of the babies and keeps them in the den for about two months. When they are big enough, they go out with their mother to find food. Their mother teaches them how to find food and how to take care of themselves so that one day they can go out and find homes of their own.

How are raccoons and people alike and different?

Classifying; recording information

Encourage the children to talk to you about how raccoons and people are alike and different. Typical responses might be:

Alike:

- ☐ We both drink milk from our mothers.
- ☐ We both breathe air with lungs.
- ☐ We both like to eat fruits, nuts, and berries.

Different:

- ☐ Raccoons have tails.
- ☐ Raccoons live in trees.
- ☐ Raccoons eat during the night.

Let's create a mother raccoon and her babies

Applying knowledge; following directions; creating a model; manipulating materials; measuring with a ruler

Mother Raccoon and her babies are fun to make from a grocery bag and a paper plate (Fig. 3-1).

Look for the hint that makes the finger puppets easy to glue in Step 7.

What you need

- ☐ 9-inch white paper plate
- ☐ Crayons
- ☐ Paper punch
- ☐ Stapler
- ☐ Ruler
- ☐ Scissors
- ☐ White glue
- ☐ Large brown paper grocery bag
- ☐ Brown, black, and white construction paper

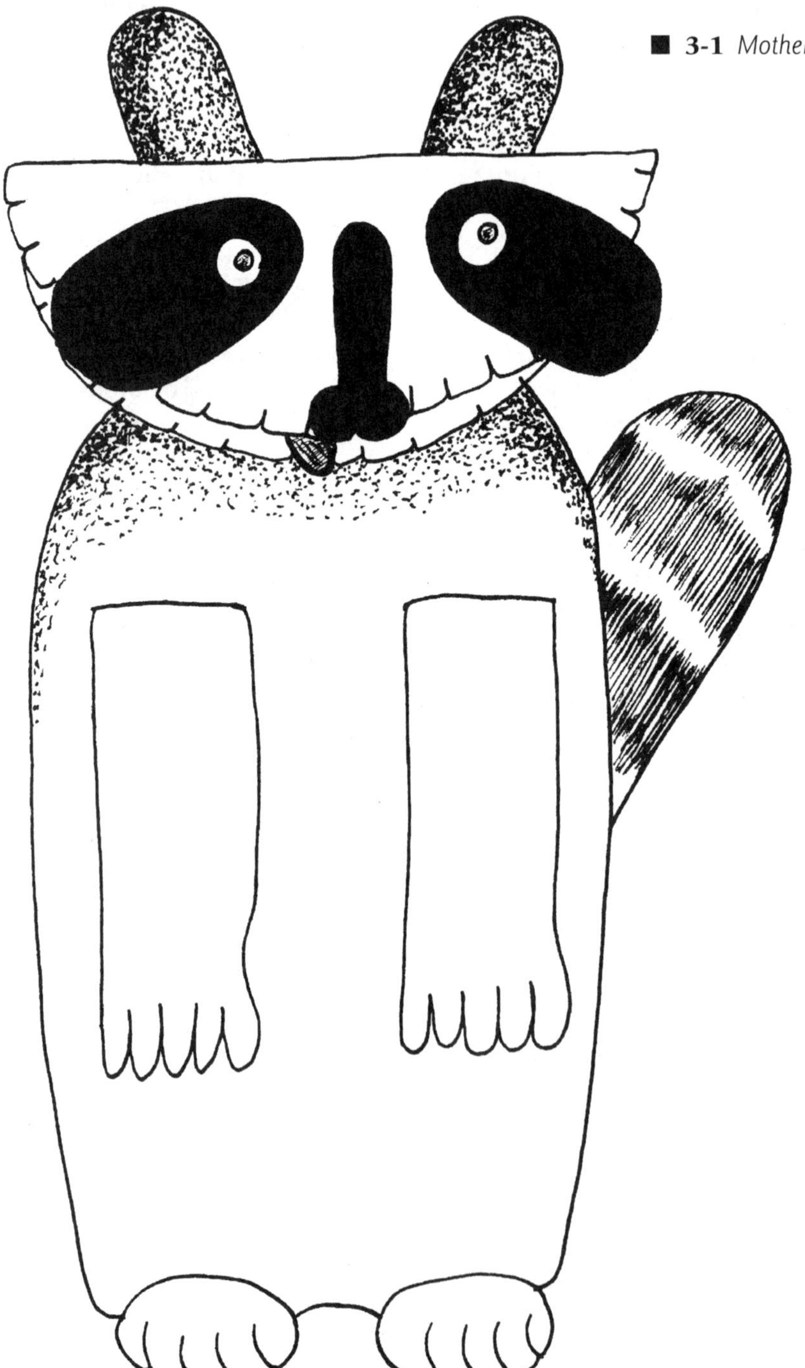

■ **3-1** *Mother raccoon.*

Directions

1. Create a raccoon head by coloring the bottom of the paper plate brown. Fold the plate in half with the coloring on the outside.
2. Use the patterns in Fig. 3-2 to trace and cut out two black oval eye masks, two brown circle eyes, two brown ears, and a black nose. To make a raccoon face, look at Fig. 3-1 as a guide for gluing these pieces onto the plate. Add a white paper punch dot to each eye.

■ **3-2** *The patterns for the mother raccoon.*

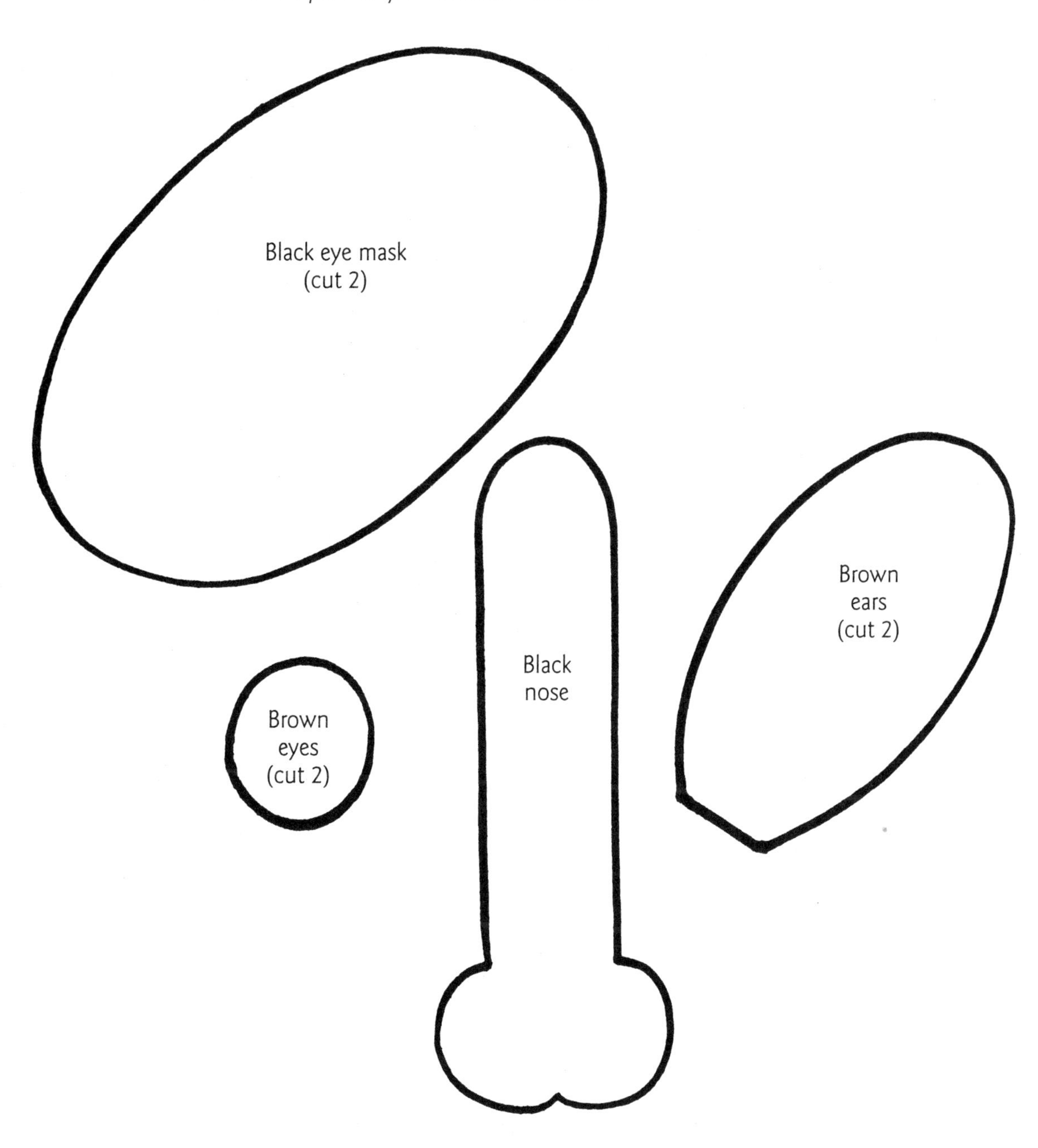

3. The remainder of the raccoon is made of pieces cut from a grocery bag. To create the raccoon body, use the ruler to measure and draw a rectangle $10\frac{1}{2}$ by 14 inches on the bag. Cut it out. Round off the corners with your scissors. Staple the paper-plate head onto the body.
4. To make the front legs, measure and draw two rectangles 3 by $6\frac{3}{4}$ inches on the grocery bag. Cut them out. Draw and cut five fingers on the end of each leg. Make the back paws from two ovals $2\frac{1}{2}$ by 4 inches, or use the eye mask as a pattern. Draw lines to show the five toes. Glue the front legs and back paws on the front of her body.
5. Design a raccoon tail from a rectangle $3\frac{1}{4}$ by $11\frac{3}{4}$ inches. Round the corners of the rectangle with scissors, and color it with black crayon stripes. Glue the tail to the back of her body.
6. Use grocery-bag paper to make raccoon-baby finger puppets as shown in Fig. 3-3. For each raccoon puppet, measure and cut out a 3-inch paper square.

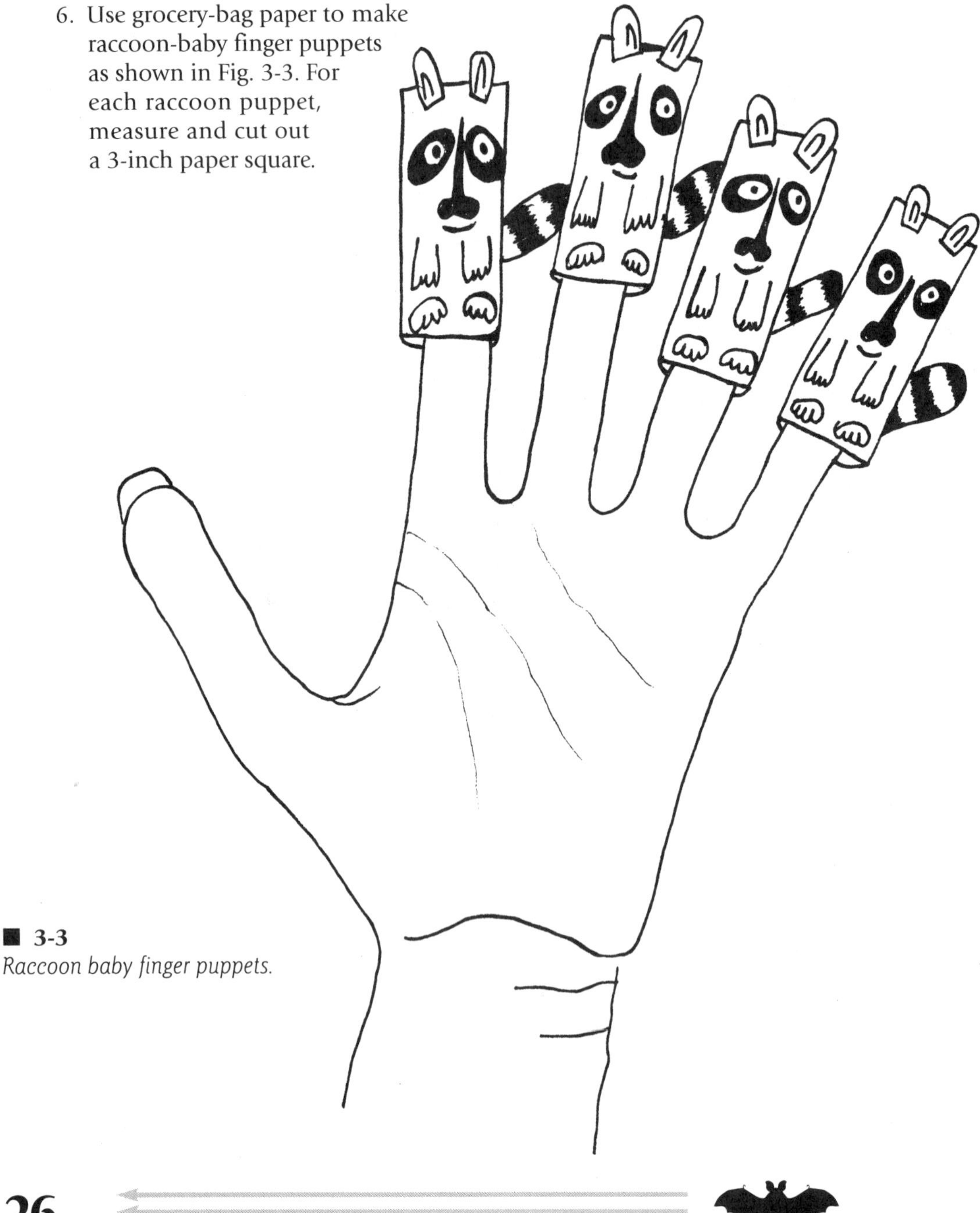

■ **3-3**
Raccoon baby finger puppets.

7. Glue each square into a cylinder shape that fits easily over your finger. Here's a hint to make gluing easy: Wrap the squares tightly around a pencil, and then remove the pencil before you glue.
8. Draw a black raccoon nose and mask on the puppets. Add brown eyes made with a paper punch. Glue on little brown paper ears and striped tails. Let the glue dry completely.

Owl

An owl is a type of bird that sleeps during the day and hunts for food at night. Most birds cannot see well enough to find food at night, but owls have very keen senses of sight and hearing and are excellent hunters at night. Owls can see as well by the light from the moon and stars as we can see by the light of the sun.

An owl is a bird. Birds belong to a group of animals called vertebrates, which means that they have backbones. Birds are warm-blooded, which means their body temperature always remains about the same regardless of the temperature of their surroundings. Baby birds hatch from eggs. Birds' bodies are covered with three kinds of feathers. *Flight feathers* are in their wings and tails to help them fly. *Downy feathers* are small and soft to keep them warm. *Body feathers* cover the rest of their body.

Birds do not have teeth. They have beaks that are used for getting food and protecting themselves. Most birds have a voice, which they use for calling and singing.

Birds have two eyes, one on each side of their head or, in the case of an owl, on the front of the face. Birds have an ear on each side of their head. They have an outer ear (covered with feathers) and an inner ear. They can hear very well.

Birds breathe in air through their mouth and nostrils (two holes in their beak) to their lungs. Some of the air travels to air sacs between different parts of their bodies. This air cools the inside of the bird.

Since birds do not have teeth, they cannot chew their food. They cut it up with their beaks or swallow it whole.

Birds have a heart, a brain, and many other organs in its body just like we do.

☞ ***Look at pictures of owls***

Finding information

Gather many different reference materials together so that the children can look at pictures of owls while you share information.

Owls' bodies are covered with feathers. Some owls even have feathers on their legs. The feathers on their legs protect them from getting bitten by other animals. There may be three or four different colors of feathers on an owl's body. Most owls are combinations of brown, black, gray, and white. The feathers of an owl's wings have very soft edges so that an owl is able to fly very quietly.

☞ *Softness*

Observing; designing investigations; drawing conclusions

Give each child a piece of writing paper and a piece of bathroom tissue or facial tissue. Let the children hold the writing paper in their left hand and the tissue in their right hand. Let the children run and flap their arms like birds. Which piece of paper is quieter?

Owls have very powerful toes on each of their two feet. These toes have long, curved, sharp talons or claws.

Owls have large eyes in the front of their heads. Their eyes are either brown or yellow. They have eyelids that close over their eyes. Owls are able to see very well at night (better than they can see in the daytime). Owls can see much better than humans can. People can roll their eyes up and down and left and right, but an owl cannot do this. Their eyes are tube-shaped instead of ball-shaped like ours. An owl must turn its head to see something on either side. An owl must raise or lower its head to look up or down.

☞ *Seeing like an owl*

Communicating information

Let the children pretend their eyes are like the eyes of an owl and they have to turn their head to see something beside, above, or below them. Call out the name of an object and let the children pretend to be owls as they look at it.

> **Nighttime Note** *Owls have more neck vertebrae than most birds. They can turn their heads 180 degrees to the right or left.*

Owls can hear very well. Most owls' ears are located behind their facial disks. Facial disks are the feathers on the face that give an owl its own special appearance. Their main function is to trap sounds and carry them directly into the ears. Some owls can hear animals that are half a mile away. Some owls can hear animals that are underground. Some owls can hear insects moving through grass.

Some owls have tufts of feathers on their heads that look like ears. These are not ears. They just make the owl look different from other owls.

☞ *Hearing like an owl*

Communicating information; applying knowledge

The children should be sitting in a group on the floor. One person at a time will be selected to be the "owl." The owl needs a flashlight. The owl leaves the room for a few minutes. While he or she is away, the teacher chooses one child to make a certain noise (for example, the child could clap softly, play a musical instrument like a triangle or rhythm sticks, or make a clicking sound with his or her mouth). Turn the lights off to simulate nighttime. Let the owl come back into the room and try to find the person making the sound by shining the flashlight all around and using his or her sense of hearing to find the sound.

> ***Nighttime Note*** *It is very difficult for young children to understand the predator and prey relationship. Handle this topic with care and use your judgment about how to discuss the topic. I try to help children realize that some animals must eat other animals to survive. Part of some animals' lives is to serve as food for other animals. Owls are very skillful hunters and catch their prey instantly. The animal that is caught never knows what happened because it is killed so quickly. Owls eat rats, mice, rabbits, squirrels, skunks, reptiles, amphibians, insects, and spiders.*

There are many types of owls. Many people think that owls make a sound like "Whoooooooooo," but each kind of owl has its own call. A barn owl may make a sound like "shreee" or "kksssch." The smallest owl in North America is the elf owl. It is 5 to 6 inches long and makes sounds like "whi-whi-whi-whi-whi-whi" and "chewk-chewk-chewk." The great gray owl, the longest owl in North America (up to 33 inches), makes a sound like "Whoooo-oooooo-ooo Whoooo-ooo-ooo-ooo." The great horned owl makes a sound like "WHO! who-who-who WHO-WHO!" The long-eared owl's call is like "kwooo-kwooo-kwooo." The snowy owl makes a sound like "Whooo-whoooo-whoooo-AAow." The screech owl makes a sound like the soft whinny of a horse.

☞ *Coloring owls*

Observing; finding information

Let the children sit down with reference books that have pictures of owls. Look through the books with the children and help them find a barn owl, a great horned owl, a screech owl, and a snowy owl. Talk about the colors of the feathers. Duplicate the pictures of these owls from this book and let the children color them (Figs. 3-4 through 3-7 *see the following four pages*).

Let's create a barn owl paper-bag puppet

Observing and creating a model of a barn owl; following directions; manipulating materials

The barn owl has a slender body and long legs. She is yellow-brown, except for her breast feathers, which are lighter with small dark brown spots. The white feathers around her face look like a valentine heart (Fig. 3-8 on page 34).

What you need

- ☐ A brown paper lunch bag
- ☐ Brown, black, and white construction paper
- ☐ Scissors
- ☐ White glue
- ☐ Crayons

Barn Owl

■ **3-4** *The barn owl.*

Great Horned Owl

■ **3-5** *The great horned owl.*

Screech Owl

■ **3-6** *The screech owl.*

Snowy Owl

■ **3-7** *The snowy owl.*

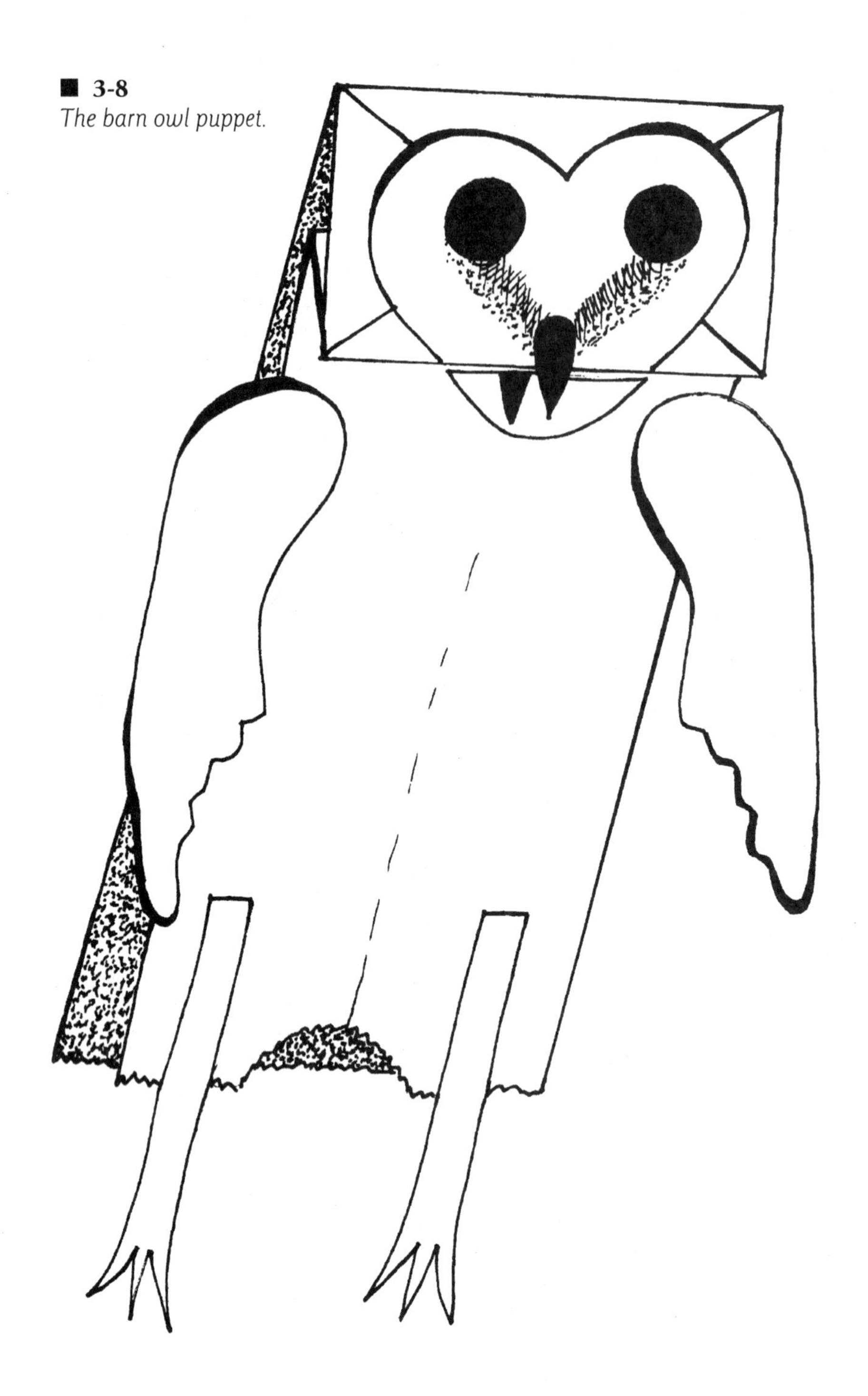

■ **3-8**
The barn owl puppet.

Directions

1. Trace and cut out the patterns in Fig. 3-9. Use them to cut two wings, two feet, and a tail from brown construction paper. Cut two beaks and eyes from black paper and the face from white paper. Cut the face into two pieces as shown on the pattern.

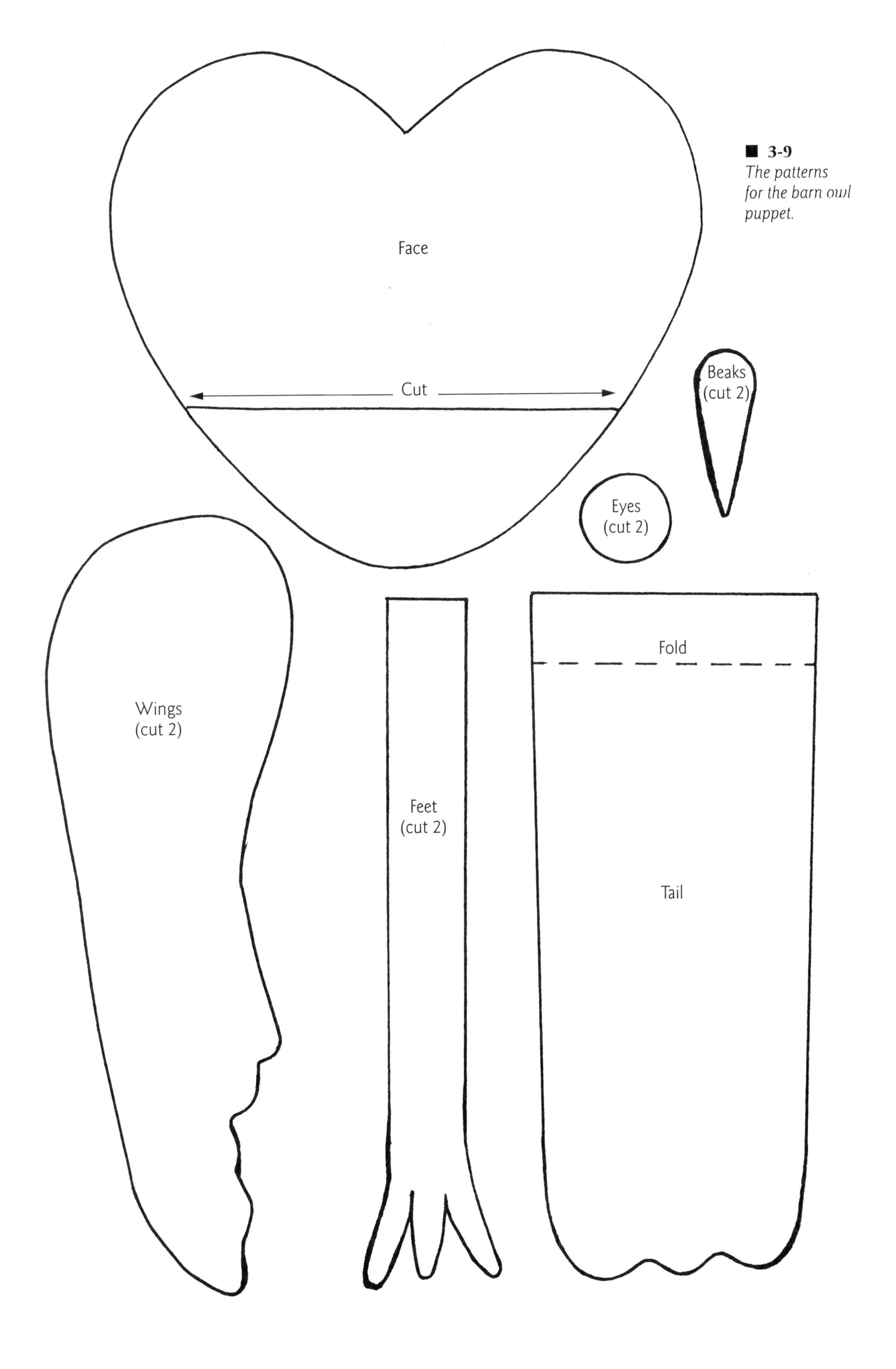

■ **3-9**
The patterns for the barn owl puppet.

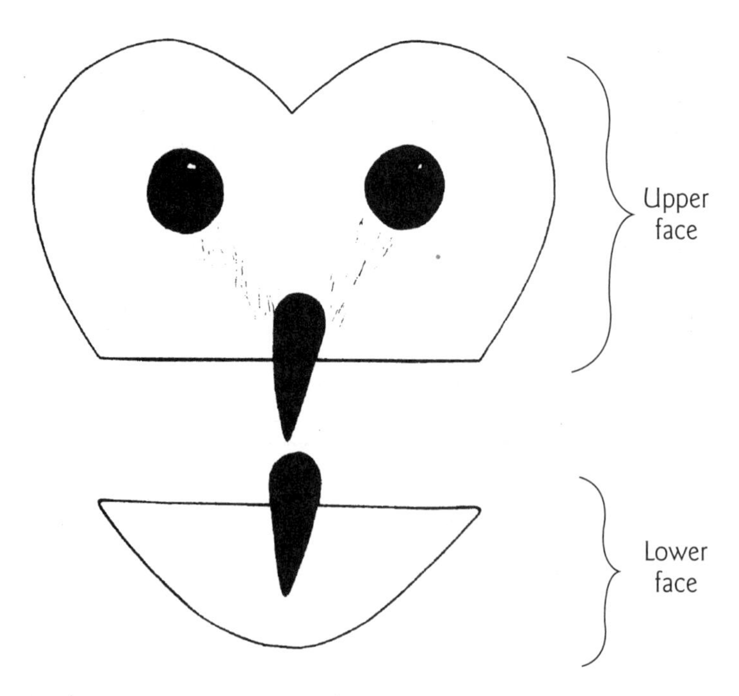

■ **3-10** *Glue the eyes and a beak on the upper face. Glue the other beak on the lower face.*

■ **3-11** *Glue the upper face on the flap of the bag. Glue the lower face below the flap.*

2. Glue the eyes and one beak onto the upper face as shown in Fig. 3-10. Glue the other beak on the lower face as shown in Fig. 3-10.
3. Glue the upper face on the bottom flap of the lunch bag. Glue the lower face below the flap so that the two face pieces line up. When you put your hand in the puppet, you can make her beak open and close (Fig. 3-11).
4. Look back at Fig. 3-8. Glue on the wings and feet. Glue the tail onto the back of your puppet. Let the glue dry.

Let's create a pine-cone great horned owl

Observing; following directions; creating a model; manipulating materials; measuring; finding information

The great horned owl is the heaviest and most powerful American owl. His "horns" are really feathers called *ear tufts* that stick up from his head. When he stretches out his wings to fly, his wingspan measures from 35 to 55 inches. Use two yard sticks to find out how wide his *wingspan* (his wings from tip to tip) might be! Go to your media center and find a book, poster, or video about the great horned owl. Then create a model from a pine cone (Fig. 3-12).

What you need

- ☐ A pine cone (about 5 inches long)
- ☐ White construction paper
- ☐ Scissors
- ☐ Crayons
- ☐ Notebook reinforcers
- ☐ A few cotton balls (cosmetic puffs)
- ☐ White glue

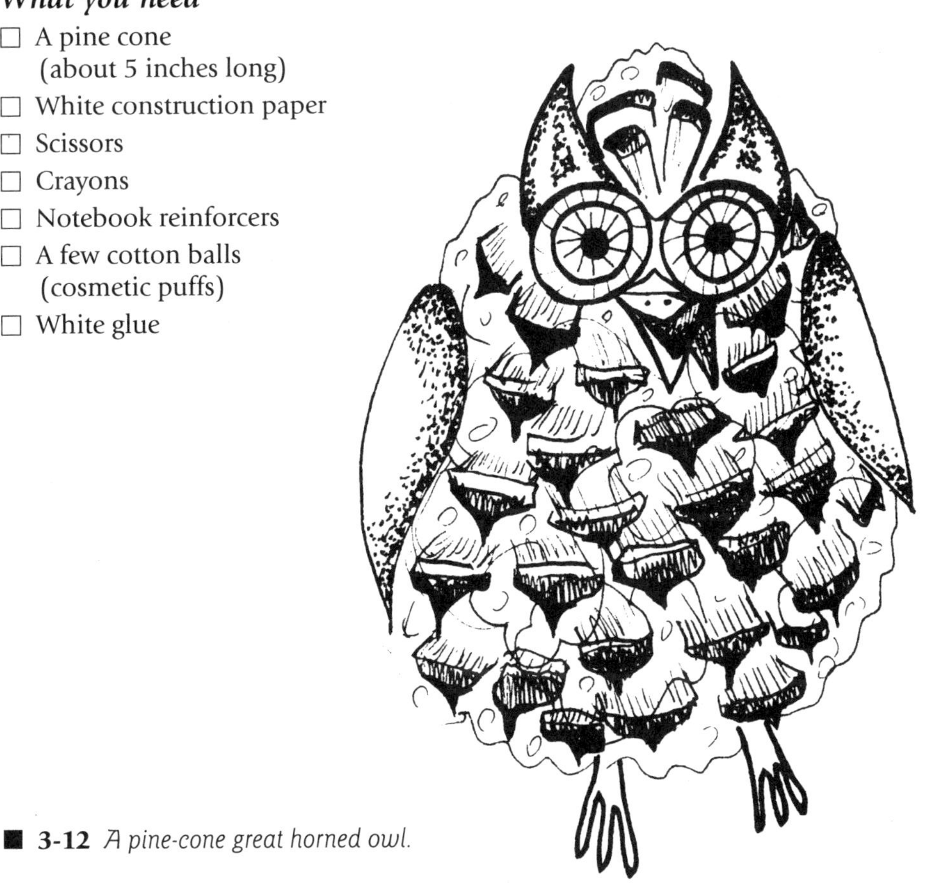

■ **3-12** *A pine-cone great horned owl.*

Directions

1. Use the patterns in Fig. 3-13 to trace feet, beak, two ear tufts, two wings, tail, and eyes onto white construction paper. Cut them out. Color the feet and beak gray. Fold the beak in half. Color the ear tufts, wings, and tail with brown and white horizontal bars (stripes). Set them aside until Step 4.

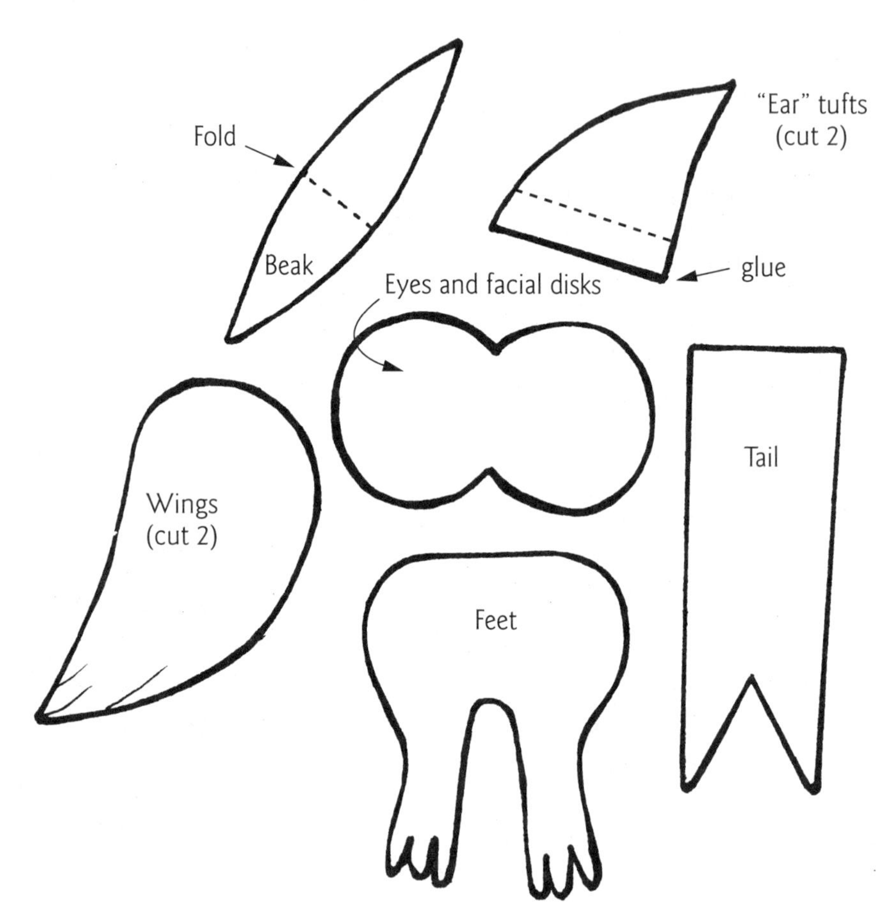

■ **3-13** *The patterns for the pine-cone great horned owl.*

2. Stick a notebook reinforcer in the middle of each eye. Color the reinforcers yellow. Color the centers black. Color the rest of the owl's facial disks with a little yellow, brown, and red (tawny). Finish the eyes by drawing a thin black line around the outside edge of each reinforcer. Set them aside until Step 4.
3. The great horned owl's feathers are brown and white with a white throat. To add whiteness to your owl, pull the cotton balls apart into small, fluffy bits and stick them in the pine cone with your fingers.
4. Look at the owl in Fig. 3-12 and glue on the facial disks, ear tufts, beak, wings, tail, and feet. Set the owl upright to dry for an hour.

Let's create a plastic-cup screech owl

Following directions; using reference materials; observing; creating a model; manipulating materials; measuring

The screech owl gets its name from the high-pitched screeching noise that it makes at dusk. It sounds similar to a horse's whinny. You can make an owl that really screeches by combining a plastic cup, a string, and a piece of wet sponge (Fig. 3-14).

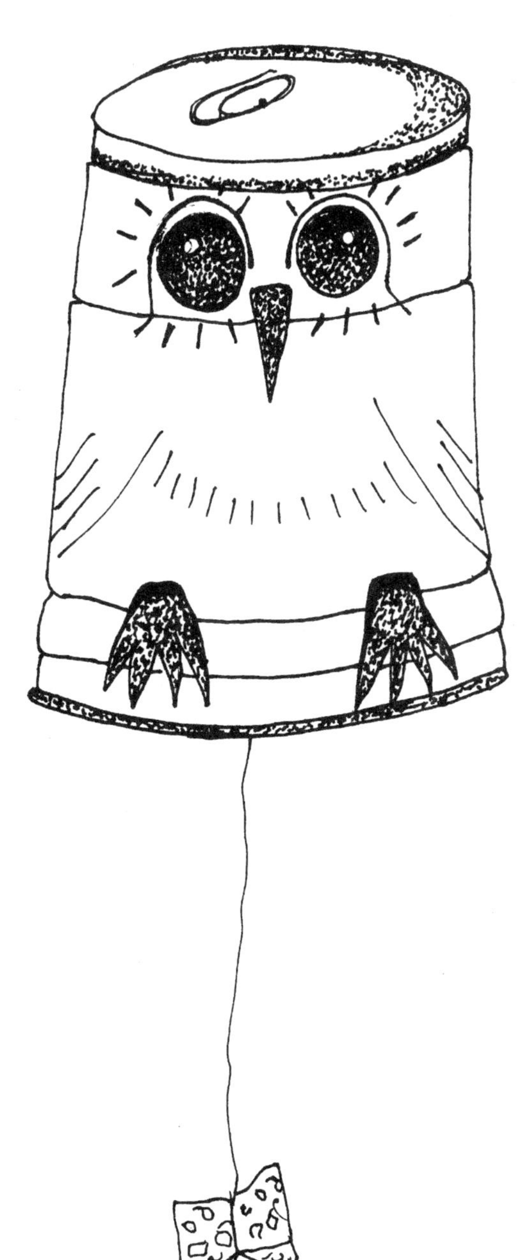

■ **3-14**
A plastic-cup screech owl.

What you need

- ☐ 16-ounce yellow disposable plastic drink cup
- ☐ Black permanent marker with a narrow tip
- ☐ Ruler
- ☐ Scissors
- ☐ 24-inch piece of embroidery thread
- ☐ Household sponge cut into a 2-inch square
- ☐ Paper clip

Directions

1. Look at the picture of a real screech owl in Fig. 3-6. To make one from a plastic cup, first ask an adult to poke a small hole in the middle of the cup's bottom.
2. Next, turn the cup upside down. Use the black marker to draw on the cup, trying to make it look like a screech owl. On our cup in Fig. 3-14, we drew the eyes, beak, wings, and tail.
3. Use a ruler and scissors to measure and cut a 24-inch piece of embroidery thread. Also measure and cut a 2-inch square of household sponge. Dip the sponge in water and then squeeze it to remove as much water as possible.
4. Tie one end of the thread very tightly around the middle of the damp sponge. Put the other end of the thread inside the cup and poke it through the hole. Tie it to a paper clip so that the sponge hangs suspended below the upside-down cup.
5. To make the owl screech, grasp the top of the string with the damp sponge and pull down with jerky motions. The wet sponge will cause the string to vibrate, and the cup will act as an amplifier to make the vibrations louder.

Let's create a snowy owl mitten puppet

Observing and creating a model of a snowy owl; following directions; manipulating materials

The snowy owl is very large and beautiful with golden yellow eyes and a rounded head. He might grow to be 24 inches tall, and his wings can spread out to 66 inches! He grows white feathers in the winter to blend in with the snow. In summer, he grows brown feathers to blend with the trees of the forest. We can create a snowy owl mitten puppet (Fig. 3-15).

What you need

- ☐ 9-x-12-inch white construction paper
- ☐ Yellow, black, and brown construction paper
- ☐ White glue
- ☐ Scissors
- ☐ Crayons
- ☐ Markers

Directions

1. In your library or media center, find some pictures of real snowy owls. With the following directions, you can create a white snowy owl puppet.
2. To make the body, curl a sheet of white construction paper into a 9-inch-tall cylinder as shown in Fig. 3-16. Overlap the edges about $1/2$ inch and glue them together. Flatten the cylinder with the glued seam down the center. Trim one end into a rounded shape like a mitten as shown in Fig. 3-17.

■ **3-15** *The snowy owl puppet.*

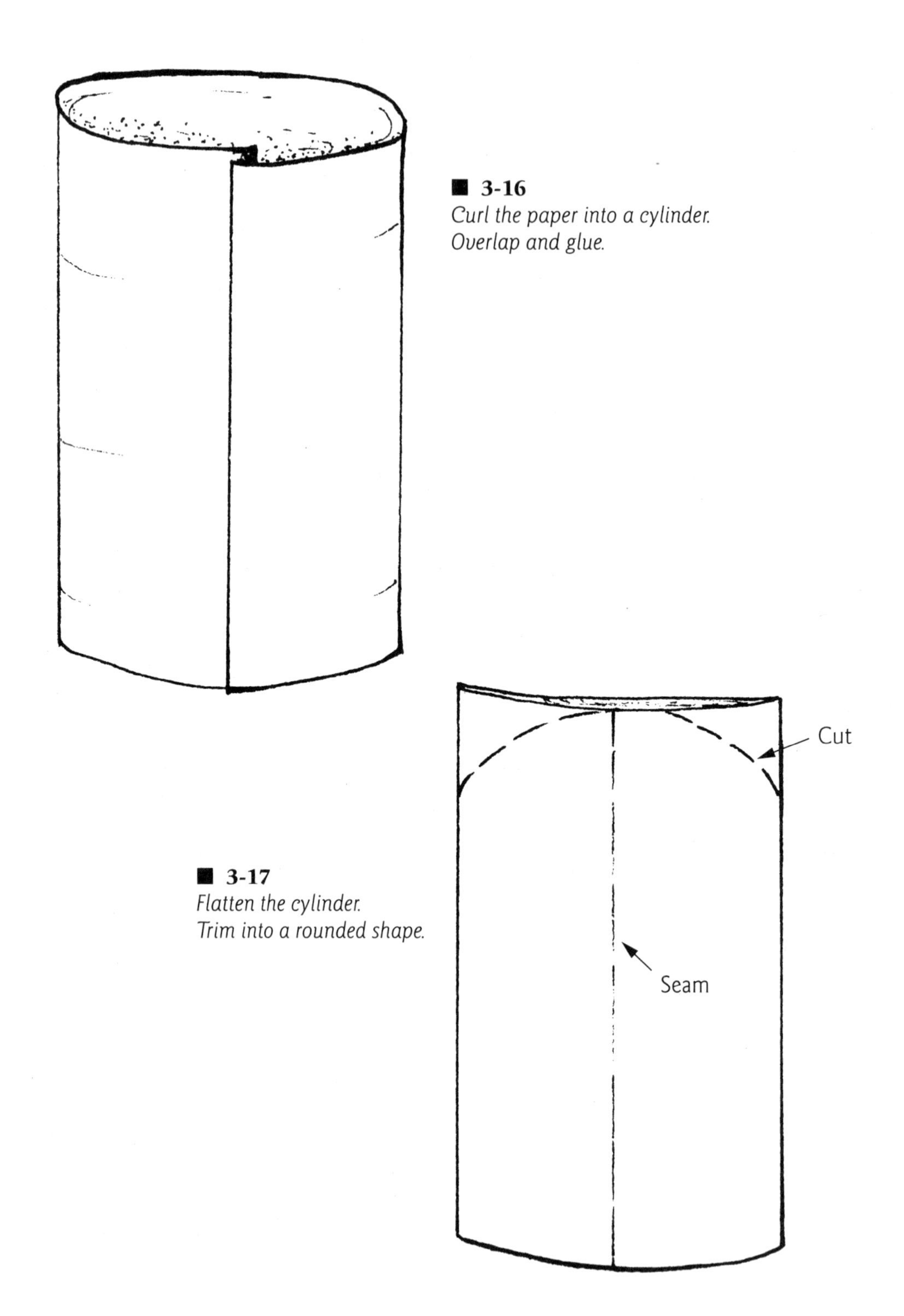

■ **3-16**
Curl the paper into a cylinder. Overlap and glue.

■ **3-17**
Flatten the cylinder. Trim into a rounded shape.

3. Use the patterns in Fig. 3-18 to trace and cut out from white construction paper two wings, two feet, two eyes, a beak, and a tail. Color the beak black. Color the eyes golden yellow with black centers. Since the feet of the snowy owl are covered with white feathers, you need only color the tips of the claws brown.

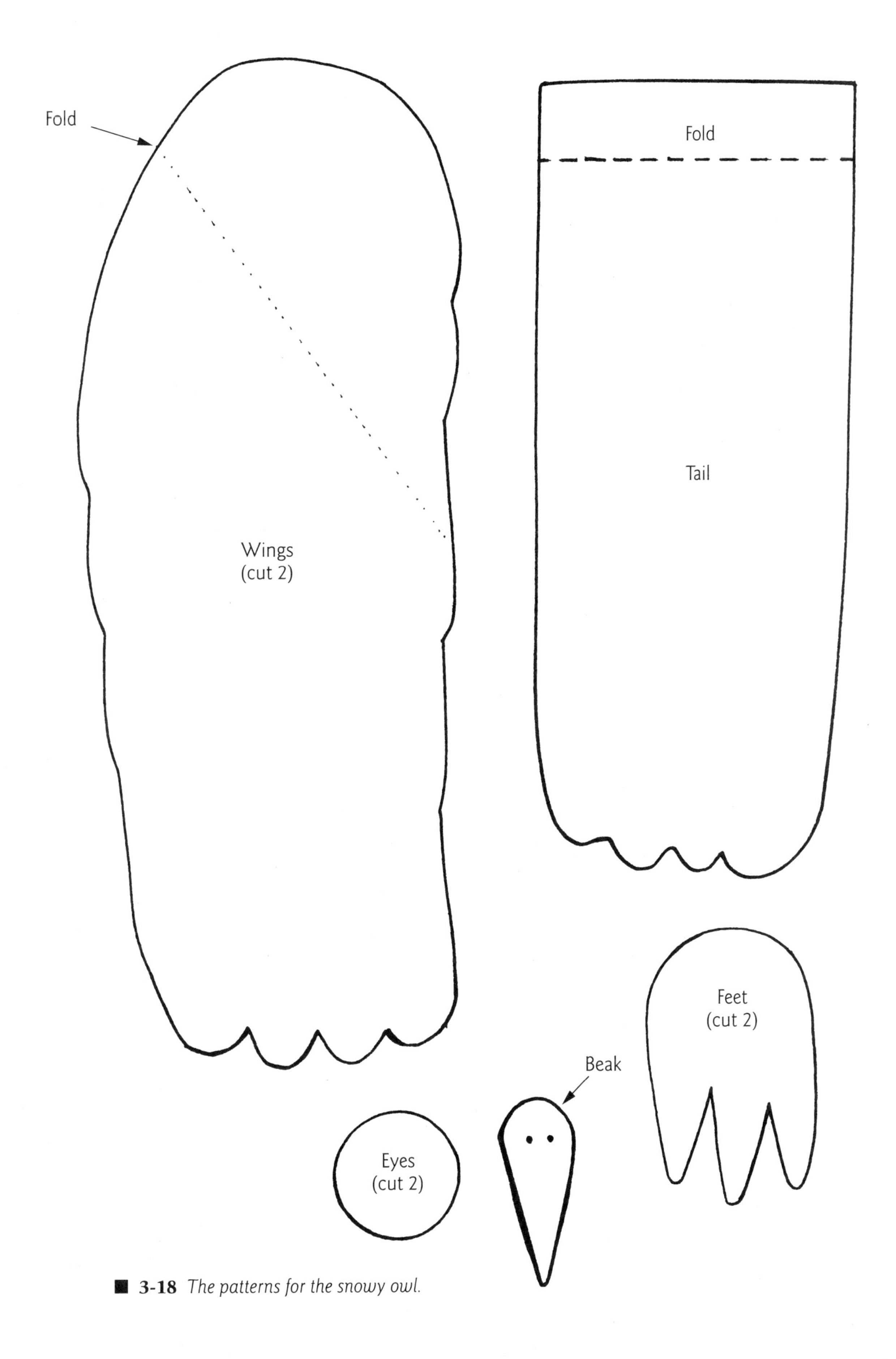

■ **3-18** *The patterns for the snowy owl.*

4. Glue the eyes and beak near the rounded edge of the mitten puppet. Glue the feet on the bottom edge. Glue the tail on the back of the puppet. Glue the wings on the sides of the puppet as shown in Fig. 3-19.

■ **3-19** *Glue the wings on the puppet.*

5. After the glue has dried, use a black marker to outline each yellow eye. Fold the wings down as marked on the pattern (Fig. 3-18).
6. To "fly" your snowy owl, put your hand inside the puppet, unfold his wings, and softly flap him up and down.

☞ ***Painting owls***

Recording information

Prepare the easel so that children can paint pictures of owls. Provide informational books with good pictures of owls. You need dark blue paper for the background. Provide brown, tan, white, black, and yellow paints for the children to use.

Bat

Bats are mammals that can fly; in fact, bats are the only mammals that can fly. Mammals are animals that have hair or fur, are warm-blooded, breathe with lungs, and their babies drink milk from their mother. Bats have a furry body. Their coloring may be black, brown, grayish, or red. Their wings are covered with smooth, flexible skin. Bats have hands with fingers and a thumb and feet with toes. The fingers support the skin of the wings. Bats' legs

are used for hanging from twigs or rocks or other objects. Most bats have one baby each year. Bats do not build nests for their babies, so the baby must hold on to its mother until it is ready to go out on its own.

> ***Nighttime Note*** *The scientific name for bats is* Chiroptera, *which means "hand-wing."*

There are more than 1,000 different species of bats. They vary in size from a bumble bee bat, which is about the size of a jelly bean, to a fruit bat, which has a wingspan of more than 5 feet. Bats live almost everywhere in the world except the Arctic and Antarctica.

Bats spend most of their lives in the darkness. During the day they live in caves, attics, or other dark parts of trees or buildings. At night they come out to find food. They use their wings to scoop up insects and toss them into their mouths.

Most kinds of bats eat insects. Some bats eat fish. Some bats eat plants. Some bats eat fruit. Most bats have sharp teeth to help them chew their food.

Some bats use their sense of sight and smell to help them find food. They also use their sense of hearing. Their ears are very sensitive. Bats use a system called *echolocation*. A bat can tell where an object or animal is by making clicking sounds that result in echoes. Using these echoes, bats can tell where things are, even trees and buildings. This ability allows the bat to fly freely in the dark without hitting anything.

☞ *Navigating by listening*

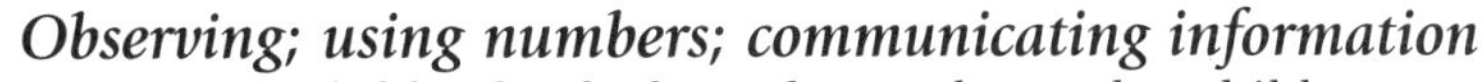

Observing; using numbers; communicating information

Choose one child to be the bat. Choose three other children to make sounds. Give each of the three children something to make a sound with, for example, wooden blocks to click together, a bell, or a drum to beat. All children should make a circle with the bat in the middle. The three children should position themselves at various places in the circle. The bat should close his or her eyes. Each child in turn should make his or her noise, and then the bat should try to walk over to the sound and touch that person. Continue playing until everyone has a turn to be the bat or make a sound.

> ***Nighttime Note*** *Did you know that bats have knees, elbows, wrists, and fingers? They also have hooked thumb claws. Bats use their thumbs and their feet to hold their food and their babies.*

Let's create a model of a common bat

Using reference materials; creating a model; following directions; manipulating materials; measuring

Like most bats, the common bat is rather small and plain-colored and eats insects at night. You can make a very good model from a toilet-paper roll

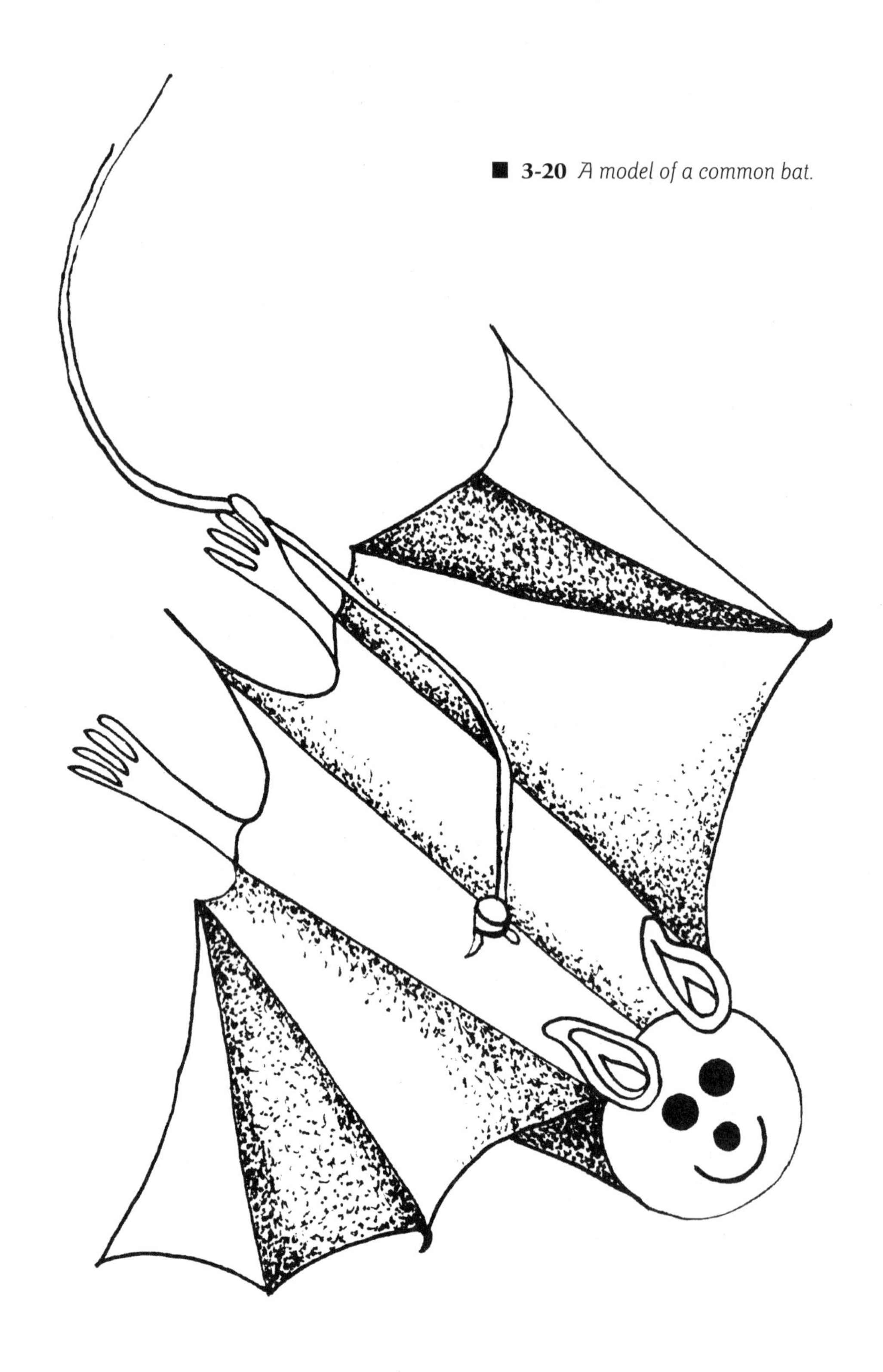

■ **3-20** *A model of a common bat.*

cylinder (Fig. 3-20). Before creating your bat model, find some pictures of real bats. If possible, read the book *Stellaluna*, the story of a baby bat that gets lost from her mother and makes new friends when she tries to act like a bird.

What you need

- ☐ Pictures of real bats
- ☐ Toilet-paper roll cylinder
- ☐ Ruler
- ☐ Scissors
- ☐ Brown and black construction paper
- ☐ White glue
- ☐ Paper punch
- ☐ Markers
- ☐ Tissue paper
- ☐ Paper fastener
- ☐ 24-inch piece of string

Directions

1. To make the body of the bat, use the ruler to draw a rectangle $4\frac{1}{2}$ by 6 inches on construction paper. Cut it out. Spread glue on the rectangle. Wrap it around the cylinder, covering the cylinder completely.
2. Use the patterns in Fig. 3-21 (shown on the next page) to trace a circle face, two ears, and two feet on brown paper. Glue the ears on the face as shown in Fig. 3-20. Add two eyes and a nose made from black circles created with a paper punch. Draw a mouth with a marker. Set aside the face and the two feet.
3. To make the wings, fold brown paper in half like a book. Use the wing pattern in Fig. 3-21 to trace wings onto the paper, with the flat side of the pattern placed on the fold. Cut through both papers, being careful not to cut the fold. Open the wings. Fold them up or down as indicated on the pattern.
4. Glue the wings to the body. Glue the feet under the wings as marked on the pattern.
5. Insert a paper fastener through the back of the bat. (An adult should use a sharp pencil to punch a hole for the paper fastener to go through.)
6. With your finger, spread glue on the front end of the cylinder. Gently place the face on the front on the bat. Let the bat dry for an hour. Then fly it around by the string.

■ **3-21** *The patterns for the common bat.*

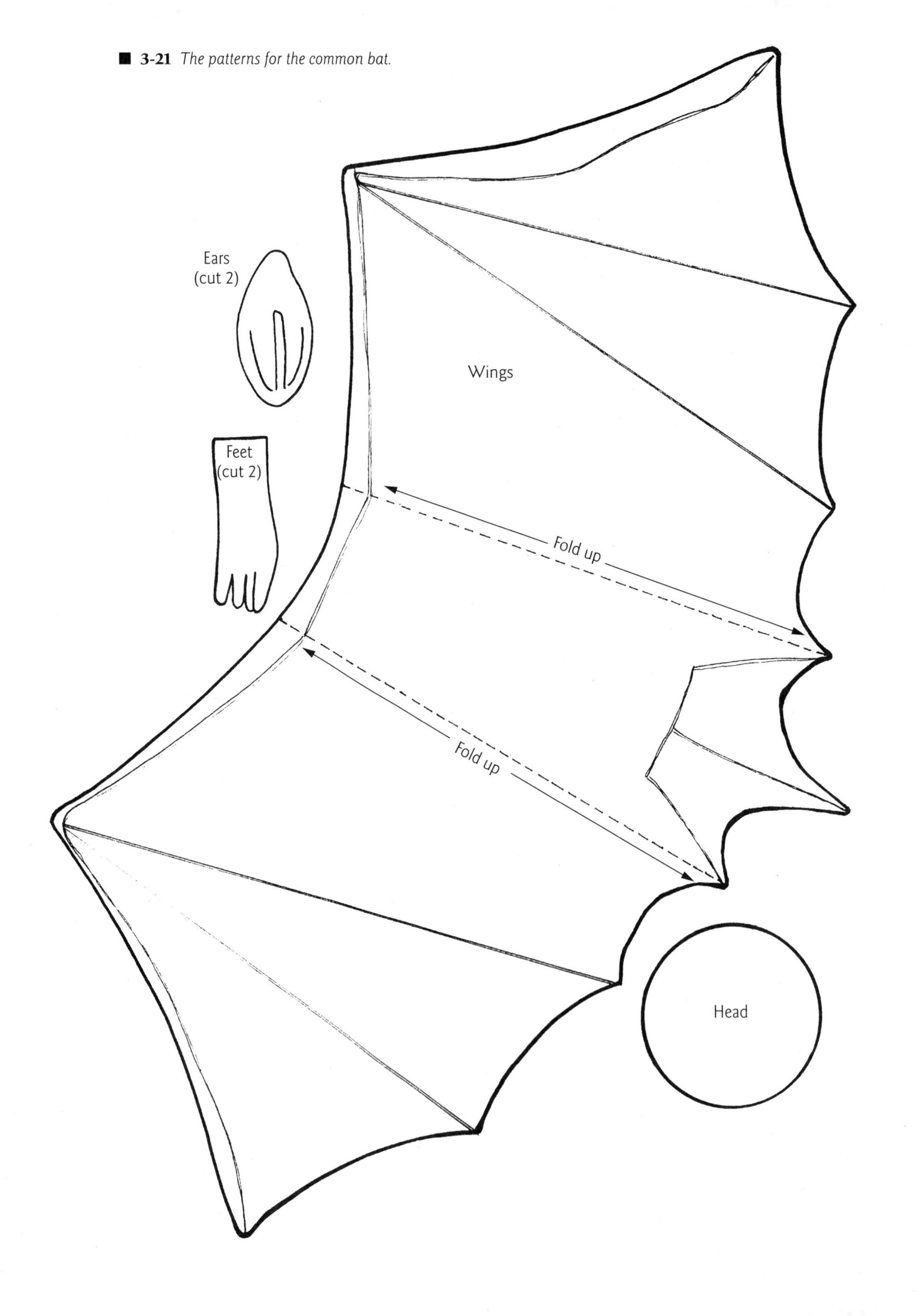

Moth

A moth is an insect. An insect has six legs, a body that is divided into three different parts, and a tough, shell-like covering. Moths go through *metamorphosis*, a process of stages of life that cause them to look different and function in different ways as they change. They change from an egg to a caterpillar, to a cocoon, and then to a moth.

Moths are relatives of butterflies. They are like butterflies in many ways, but they are different, too. Moths and butterflies have four wings. Generally, butterflies fly during the daytime and moths fly at night. Since most moths are out at night (some are out during the day), we miss seeing them. The best place to find moths is near a light outdoors. Moths are attracted to fluorescent lights. Most moths are not as colorful as butterflies. Many moths are brownish in color. The antennae of moths look like feathery rabbit ears or are smooth and tapered. When a moth is resting, it holds its wings out flat over is body.

Adult moths can be less than 1 inch across or as much as 12 inches across. An adult moth has a head, thorax, and abdomen. The four wings are attached to the thorax. Its body is covered with scales and hairs. Most moths cannot bite or chew. They use their proboscis (a tube coming from the mouth) to suck up food from flowers. Moths drink water, sip nectar, and drink sap from trees.

☞ ***Creative movement to light and music***

Observing; communicating information; applying knowledge

Tell the children that they can pretend to be moths at night. Turn the overhead lights off and turn on a small lamp in the middle of the room with plenty of space for movement. Play calm music. Let the children pretend to be moths and fly around the light. Encourage them to sip nectar from plants and to stop and rest at times. When the overhead lights come back on, it means that it is daytime, and the moths should go to sleep quietly.

Let's create a model of a moth

Observing; creating a model; following directions; manipulating materials; measuring

Most moths sleep during the day and are seen at night. This moth is glued onto a large craft stick, which are sometimes found in doctors' offices as "tongue depressors" (Fig. 3-22).

What you need

- ☐ 6-inch craft stick (or tongue depressor)
- ☐ Markers
- ☐ Wiggle eyes
- ☐ Construction paper
- ☐ Scissors
- ☐ White glue

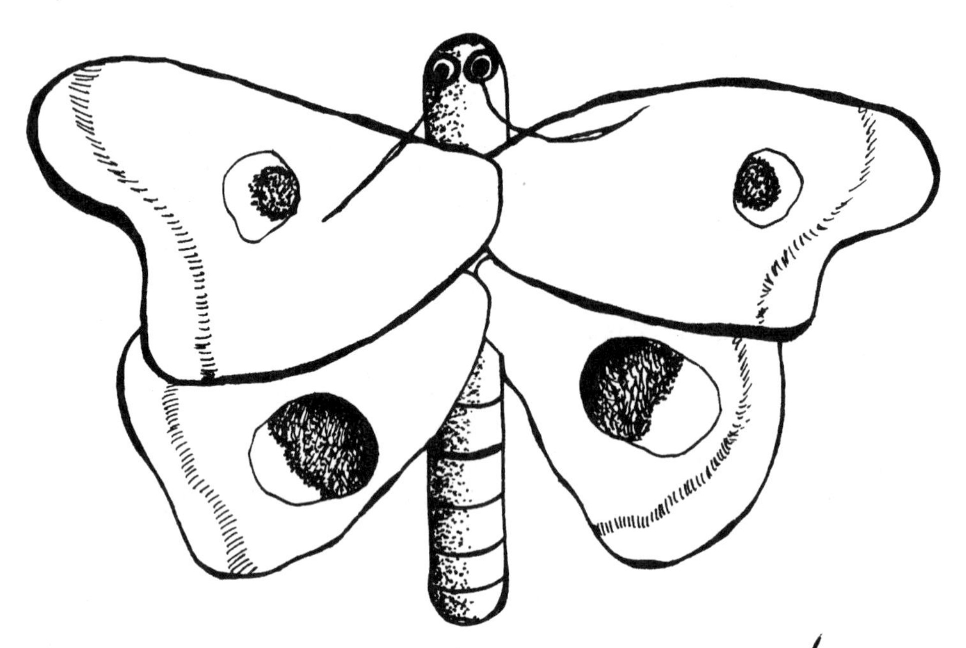

■ **3-22** *A moth.*

Directions

1. Use markers to draw a moth head, thorax, and abdomen on the craft stick as shown in Fig. 3-23. Glue on wiggle eyes. Cut two thin paper strips for antennae and glue them above the eyes.

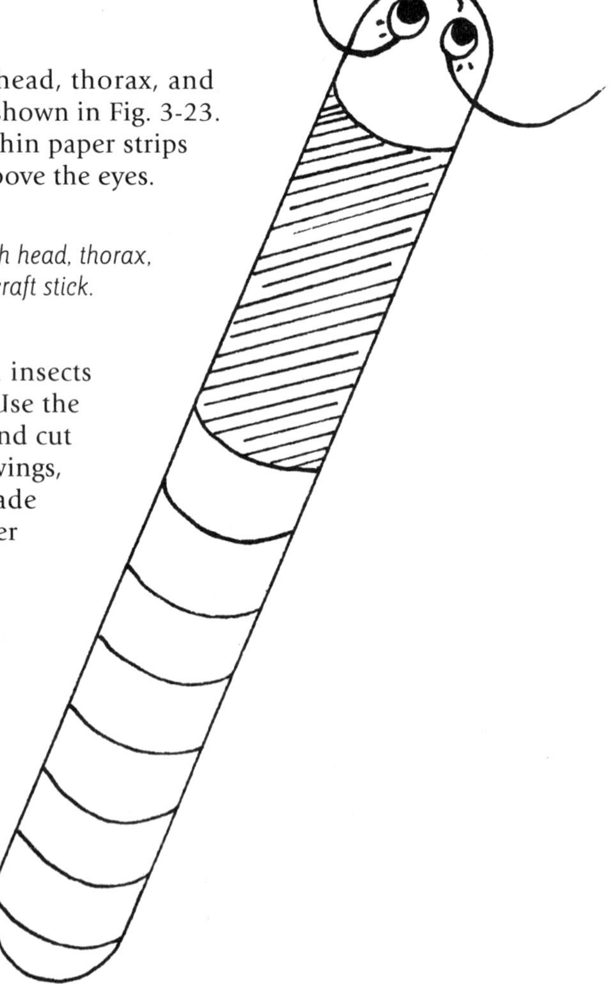

■ **3-23** *Draw a moth head, thorax, and abdomen on the craft stick.*

2. The moth is an insect, and all insects have four wings and six legs. Use the patterns in Fig. 3-24 to trace and cut out two fore wings, two hind wings, and the legs. These can be made of any color construction paper you wish to use.
3. Glue the wings on top of the moth's thorax.
4. Glue the legs on the bottom of the thorax.
5. After the glue dries, you can "fly" your moth by holding its abdomen and gently flapping it up and down.

■ **3-24** *The patterns for the moth.*

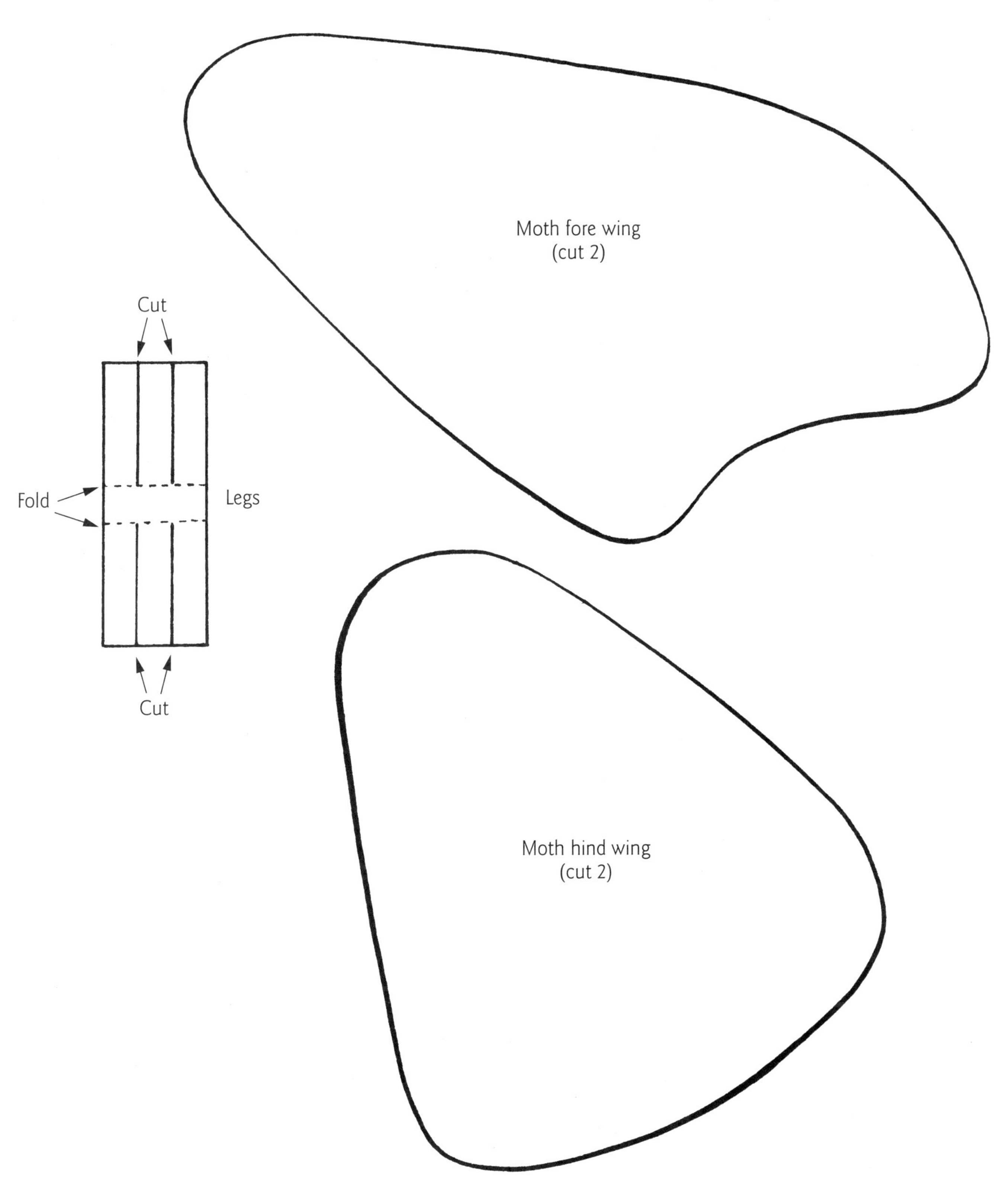

Firefly

Fireflies are insects (beetles) that glow in the dark. Insects have six legs, three body parts, and a shell-like outer covering. Most fireflies have wings, but some do not. Fireflies are sometimes called lightning bugs. Fireflies like to live where the weather is warm, and they are frequently seen during the summer.

The light that fireflies produce helps fireflies find each other. They call to each other with twinkling lights. Chemicals in the firefly's abdomen cause the lights. Some fireflies give off yellow light, and some give off red and green lights. In some parts of the world, people wear fireflies on their clothing to provide light at night.

☞ *Colored lights game*

Observing; following directions; manipulating materials; using numbers; communicating information; applying knowledge

Get six flashlights. Leave two as they are. Cover the front of two with green tissue paper or cellophane, and cover the front of two with red tissue paper or cellophane. These flashlights provide three kinds of light: white or yellow, green, and red. Position six children in different places in the room. Turn the lights in the room off. Give each child a flashlight. Let them blink their lights on and off. The object of the game is for partners with the same color of light to find each other and get together. The game can end when all partners are together.

Let's create a model of a firefly

Observing; creating a model; following directions; manipulating materials; measuring

Creating a model of a firefly (Fig. 3-25) is a terrific way for a student to learn about night-flying insects, to illustrate a story, to develop eye-hand coordination skills, to build language skills, and to increase positive self-esteem.

What you need

- ☐ Ruler
- ☐ Brown, yellow, and black construction paper
- ☐ Tissue paper
- ☐ Paper punch
- ☐ White glue
- ☐ Markers

Directions

1. Use a ruler and scissors to measure and cut two 2-by-9-inch brown strips and one 2-by-9-inch yellow strip. Glue the strips together to make a paper chain. The first brown strip is the firefly's head. The second brown strip is the thorax. The yellow strip is the firefly's abdomen (Fig. 3-26).

■ **3-25** *A firefly model.*

■ **3-26** *Make a paper chain.*

2. The firefly is an insect, and all insects have four wings. Use the pattern in Fig. 3-27 to trace and cut two wings from black construction paper and two more wings from tissue paper.

■ **3-27** *The patterns for the firefly wings, eyes, and legs.*

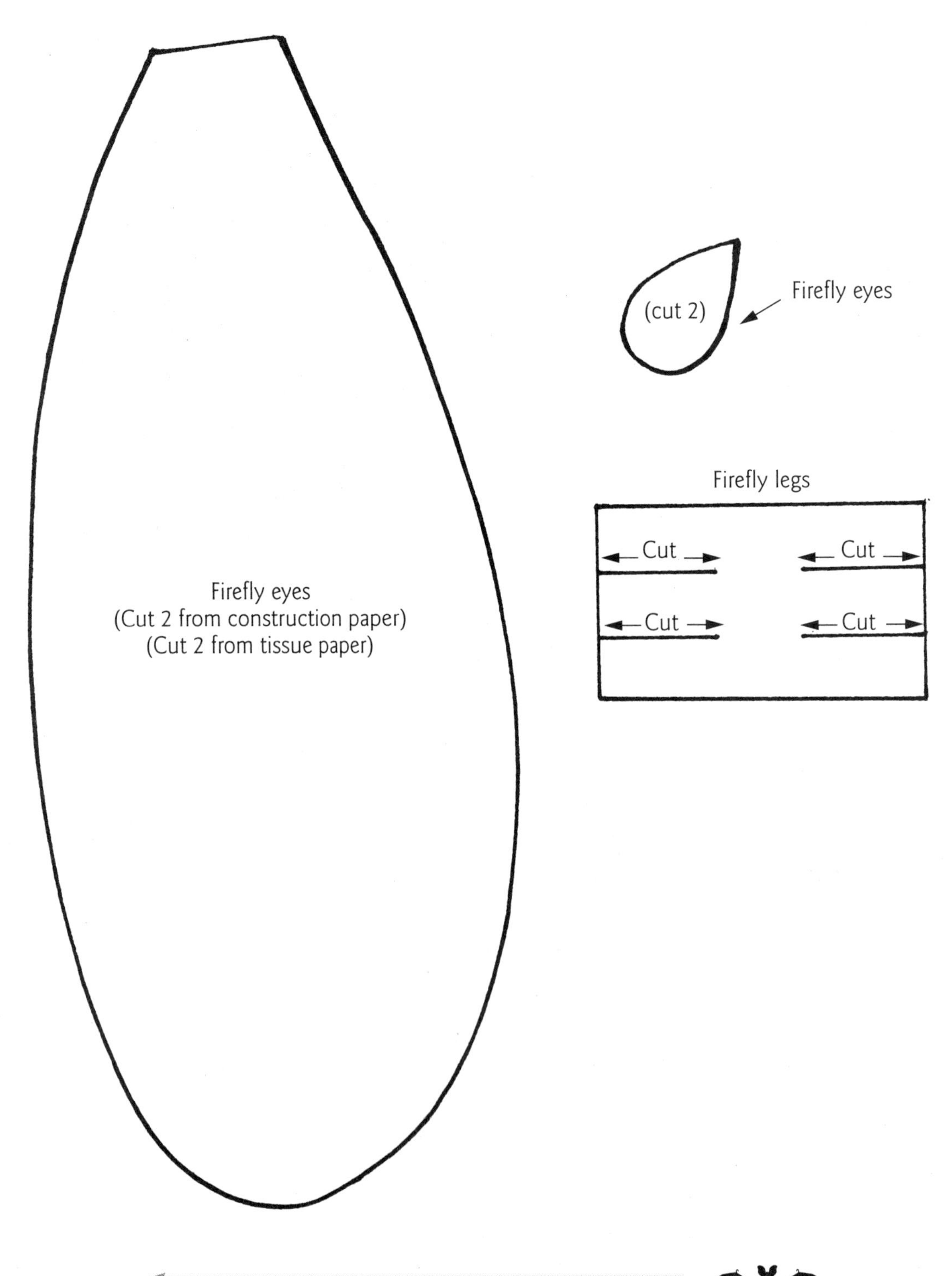

3. Glue a tissue paper wing loosely under each construction paper wing (Fig. 3-28). Then glue the combined wings on top of the firefly's head.

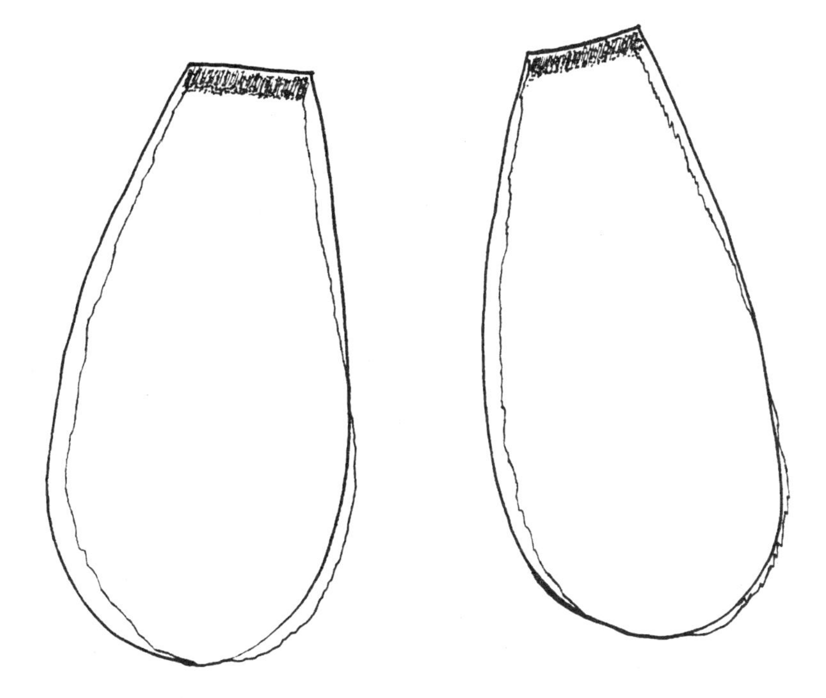

■ **3-28** *Glue a tissue-paper wing under each construction-paper wing.*

4. Add glue to the top of the abdomen and stick the wings on there, too.
5. Use the patterns in Fig. 3-27 to cut out black eyes and legs. Glue the eyes to the face. Add white paper-punch eyeballs.
6. Glue the legs under the head. Bend down the six legs.
7. Glue two thin strips of black paper above the eyes for antennae. Draw eyebrows and a mouth with a marker.

Extension activities

☐ After creating the firefly model, children ages 6 and up can use reference materials to find out about fireflies and to discover information about other nocturnal insects such as crickets and moths.

☐ Children can write, illustrate, and act out an imaginary story about a firefly's adventures.

Let's create puffy-paint night crawlers

Following directions; creating a model; manipulating materials; measuring

Night crawlers are large earthworms that come out of their underground tunnels at night. You can create some very real-looking night crawlers by mixing your own puffy paint and squirting it onto coffee grounds that look like dirt.

What you need

- ☐ Corrugated cardboard rectangle (about 8 by 10 inches)
- ☐ White glue
- ☐ Coffee grounds (used coffee grounds that have been dried)
- ☐ A funnel
- ☐ Plastic squirt bottle with removable lid—the type used for catsup or mustard
- ☐ 1/3 cup salt
- ☐ 1/3 cup flour
- ☐ 1/4 cup water
- ☐ Brown washable paint
- ☐ 1 tablespoon water
- ☐ 1/4 cup salt
- ☐ 1/4 cup flour

Directions

1. Spread white glue thinly on the corrugated cardboard. Sift on coffee grounds to make the board look like it's covered with dirt. Set it aside to dry.
2. Remove the lid from the squirt bottle. Make your own puffy paint by measuring 1/3 cup salt and 1/3 cup flour into the bottle. (Use the funnel to help you.) Shake the bottle to mix them. Add 1/4 cup water and a squirt of brown paint. Put the lid on the bottle and shake it hard.
3. Remove the lid. Use the funnel again to add 1 tablespoon water, 1/4 cup salt, and 1/4 cup flour. Replace the lid. Shake this mixture very hard.
4. Squirt out night crawlers onto the prepared "dirt" (Fig. 3-29). Let them dry overnight.

Let's create night-crawler paintings

Observing; following directions; manipulating materials

Dip night crawlers in washable paint, and let them crawl on a piece of paper. Their tracks make very interesting "art."

What you need

- ☐ Night crawlers (can be purchased from a pet store or bait shop)
- ☐ Tempera washable paint
- ☐ Foam meat tray from the grocery store
- ☐ Craft sticks
- ☐ White construction paper

Directions

1. Pour a small amount of washable paint into a clean meat tray. Stir it with a craft stick.
2. First lay a night crawler in the paint, and then put it on your construction paper. It will crawl around, leaving paint trails (Fig. 3-30).

■ **3-29** *Puffy-paint night crawlers.*

■ **3-30** *Night crawler paintings.*

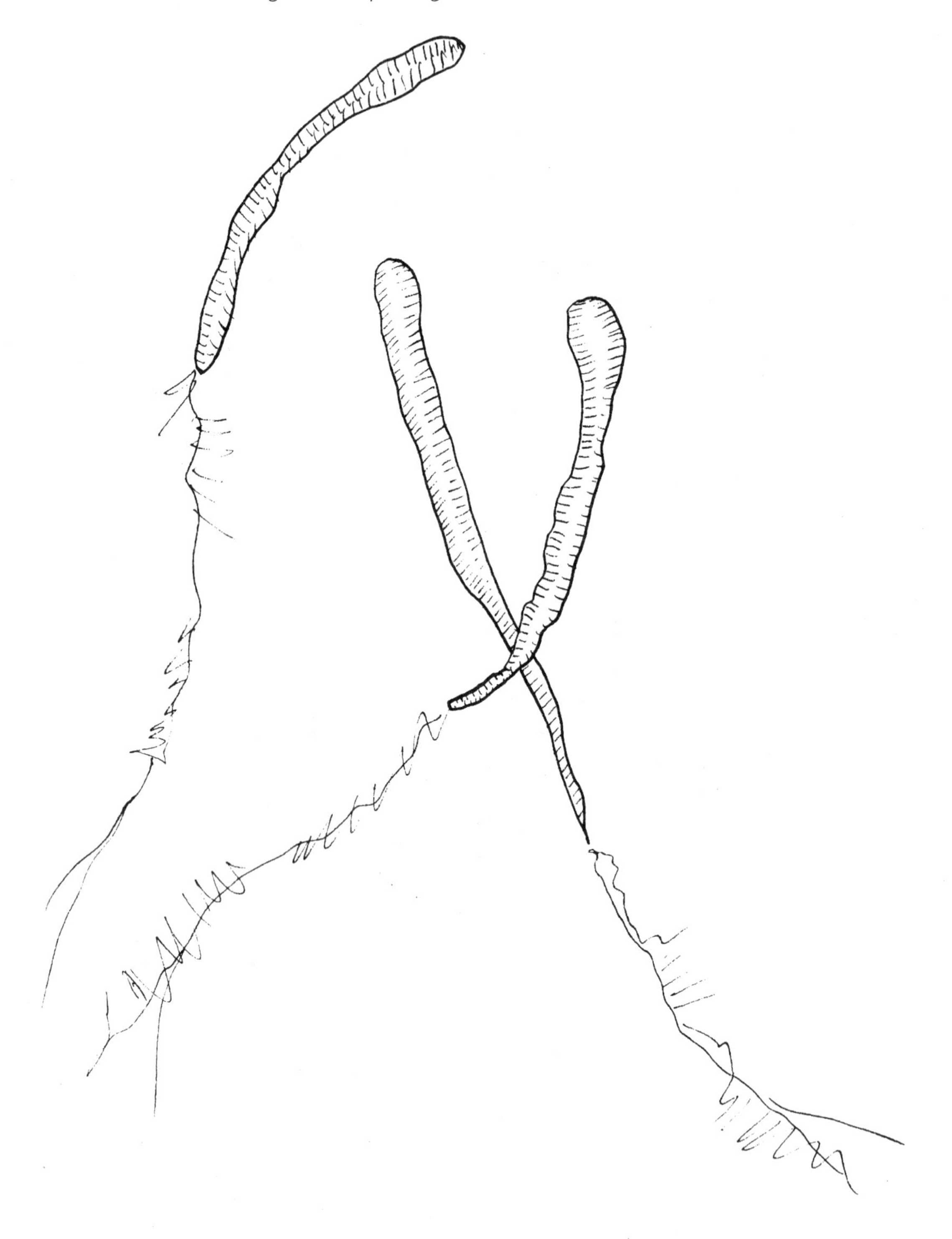

3. When you have finished painting, wash off the night crawlers with cool water, and release them outside in a patch of shaded dirt or grass.

Let's create chocolate night crawlers

Following directions; manipulating materials; measuring
Chocolate night crawlers are fun to make, and they taste delicious!

What you need

- ☐ 12 ounces (2 cups) chocolate chips
- ☐ 6 ounces ($3^1/_2$ cups) chow mein noodles
- ☐ Large microwavable bowl
- ☐ Wooden spoon
- ☐ Two or more cookie sheets covered with waxed paper

Directions

1. Measure the chocolate chips in the bowl. Microwave on high for three minutes. Stir the melted chips with a spoon until smooth.
2. Add the chow mein noodles. Stir to coat them with chocolate. Be careful not to break the noodles.
3. Spoon the noodles onto the waxed paper-covered cookie sheets. Use your fingers to separate them into chocolate night crawlers.
4. Refrigerate the cookie sheets for 30 minutes until the chocolate hardens. Eat the night crawlers.

Beaver

A beaver is a mammal. Mammals have backbones, hair or fur, are warm-blooded, breathe with lungs, and their babies are born alive and need milk from their mothers. Beavers are furry animals. Their fur is either dark or light brown. They have an interesting tail that is flat and looks like a paddle. This special tail helps beavers swim. It helps the beaver steer itself in the water. Sometimes beavers slap their tails against the water to make a loud noise. The noise tells other beavers that danger is near.

Beavers are known for their ability to cut down trees with their strong, sharp front teeth. Beavers have 20 teeth. Four of these teeth are for gnawing and the others are for chewing. The four teeth for gnawing are called *incisors*, and they keep growing throughout the beaver's life. When they get worn down from gnawing or cutting down trees, they just grow some more. Inside a beaver's mouth are flaps of skin that separate these front teeth from the other 16 teeth. When a beaver is gnawing on a tree, these flaps of skin keep the bits of wood from going into the beaver's throat.

Beavers are good swimmers and divers. Their back feet are webbed to help them be good swimmers. Their front legs are shorter than their back legs. Their front feet are designed for digging and getting food. The front feet have five toes with claws.

Beavers eat many parts of plants. They like to eat the bark of trees. They also like the leaves and roots of trees and bushes. Beavers like the roots of plants that live in the water too.

Beavers usually come out at night to eat and carry out their work. Beavers have families. A family is made up of the mother and father and the babies. Babies live with their parents for about two years. Then they go out and start a family of their own.

Beavers build their homes, which are called *lodges*. First, beavers cut down trees. When beavers cut down trees, they stand up and let their tails support them on the ground. They hold onto the tree with their two front feet. Then they turn their heads sideways and make cuts in the tree, chewing and pulling until the tree falls. Then the beavers gnaw the branches off the tree and gnaw off bark. Beavers carry the logs to the water and use them to build or repair their lodge. They also store some logs for food.

A lodge is made of branches, logs, and maybe rocks held together by mud. The base of the lodge is in the water, and some of it is above the water. Lodges have several entrances that are under the water. These entrances lead to areas above the water that are dry.

"Busy as a beaver"

Asking questions

Ask the children, "Have you ever heard anyone say someone is 'as busy as a beaver'? Beavers work very hard, so if anyone says this about another person, it is really a nice compliment." Let the children talk about some times that this phrase may have described them.

☞ ***Build with sticks***

Creating models; manipulating materials; applying knowledge

Ask the children to bring all kinds of small sticks and rocks from home or gather them yourself. Try to find various kinds of sticks (some that are thin and flexible, for example) that are about 6 inches or less long. Provide a small pan of water. Let the children build a lodge in the water. The rocks will be helpful in making the sticks stay down in the water (they may float otherwise). Be sure to remind the children that part of the lodge should be above the water so the beavers can stay dry.

Let's create a little beaver book

Creating a model of a beaver; following directions; manipulating materials

A beaver has small round ears, short legs, and large webbed hind feet that help it swim easily through the water. It has thick, brown fur and a flat, scaly tail that is about a foot long. You can make a little beaver book from a file folder.

What you need

- ☐ A file folder
- ☐ Scissors
- ☐ Stapler
- ☐ Paper for the book's pages
- ☐ Crayons

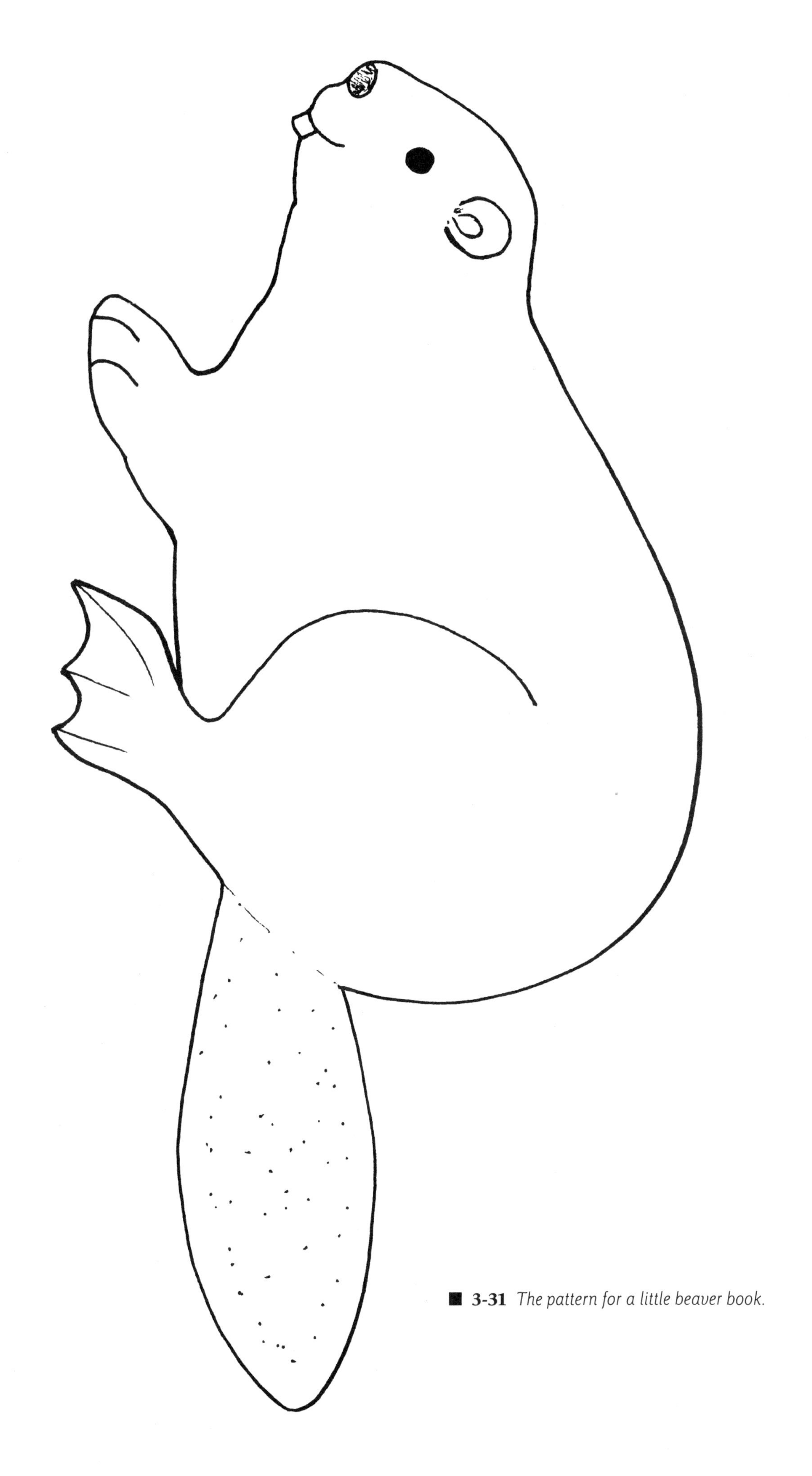

■ **3-31** *The pattern for a little beaver book.*

Directions

1. Trace and cut out the beaver pattern in Fig. 3-31. Lay the pattern on the file folder and draw around it. Cut out two beaver shapes—one for the front of the book and one for the back.
2. Use the same pattern to cut out pages for your book.
3. Color both the front and the back of the beaver. Staple the book together with the pages between the shapes.

Uses for the little beaver book

- ☐ Write a story about a beaver or about any nocturnal animal. Is it a real story or is it imaginary? Illustrate each page of your story.
- ☐ Write a factual report about a beaver. Include information on what it eats, how it swims, and where it lives.

Opossum

Opossums are mammals. Mammals have backbones, hair or fur, are warm-blooded, breathe with lungs, and their babies are born alive and need milk from their mother. Opossums are furry animals that carry their babies in the mother's pouch. Kangaroos also do this. When a baby opossum is born, it is only about a half inch long. The mother keeps her babies in a pouch on her abdomen for about two months. After the babies are able to leave the pouch, they stay near their mother for several weeks. They like to ride on her back when they go someplace.

When an opossum is grown, it may be about the size of a cat. Opossums look kind of funny because their fur is rough, their ears don't have any hair on them, and their tails have hardly any hair either.

Opossums have sharp teeth and claws. They have long toes on each foot. They can hang upside down by wrapping their tail around a limb.

Opossums hunt at night. They eat almost anything they can find, whether meat or vegetables.

"Playing possum"

Asking questions

Have you ever heard anyone say, "You're playing possum"? This is an old saying that means you are pretending to be asleep or injured. When opossums are afraid of danger, they may lie very still and not move at all. "Why do you think they would do this?"

Talk about how other animals respond to danger or just to a person being near. For example, a bird usually flies away when a person tries to come near, even if the person just wants to look at the bird. A cat may hide from a stranger. A dog may bark and run toward a stranger.

☞ *Let's play possum*

Observing; following directions; communicating information; applying knowledge

One person should be designated to pretend to take a walk through the woods. On this walk he or she will pass by different kinds of nocturnal animals. Each animal will respond in a different way.

Decide on the kinds of animals you want and which children will pretend to be these animals. Decide what action the animal will take as the person walks by. Here is a possible list:

Animal	How the animal will act
Firefly	Fly around and blink its light
Raccoon	Take the food it is eating and walk away
Opossum	Lie very still and close its eyes
Moth	Fly slowly away
Owl	Sit very still, turning its head from side to side
Bat	Fly away
Beaver	Slip into the water and go to its lodge

Let's create a pocket opossum

Applying knowledge; following directions; creating a model; manipulating materials; measuring with a ruler

The opossum is the only marsupial in North America. *Marsupial* means that the mother opossum has a pocket. She carries her babies in the pocket when they are first born.

You can make a pocket opossum from a paper bag and a paper plate (Fig. 3-32 shown on next page). Look for the hint that makes the finger puppets easy to glue in Step 9.

What you need

- ☐ 9-inch white paper plate
- ☐ Crayons
- ☐ Paper punch
- ☐ Stapler
- ☐ Ruler
- ☐ Scissors
- ☐ White glue
- ☐ Large brown paper grocery bag
- ☐ Brown, black, and white construction paper

■ **3-32** *A pocket opossum.*

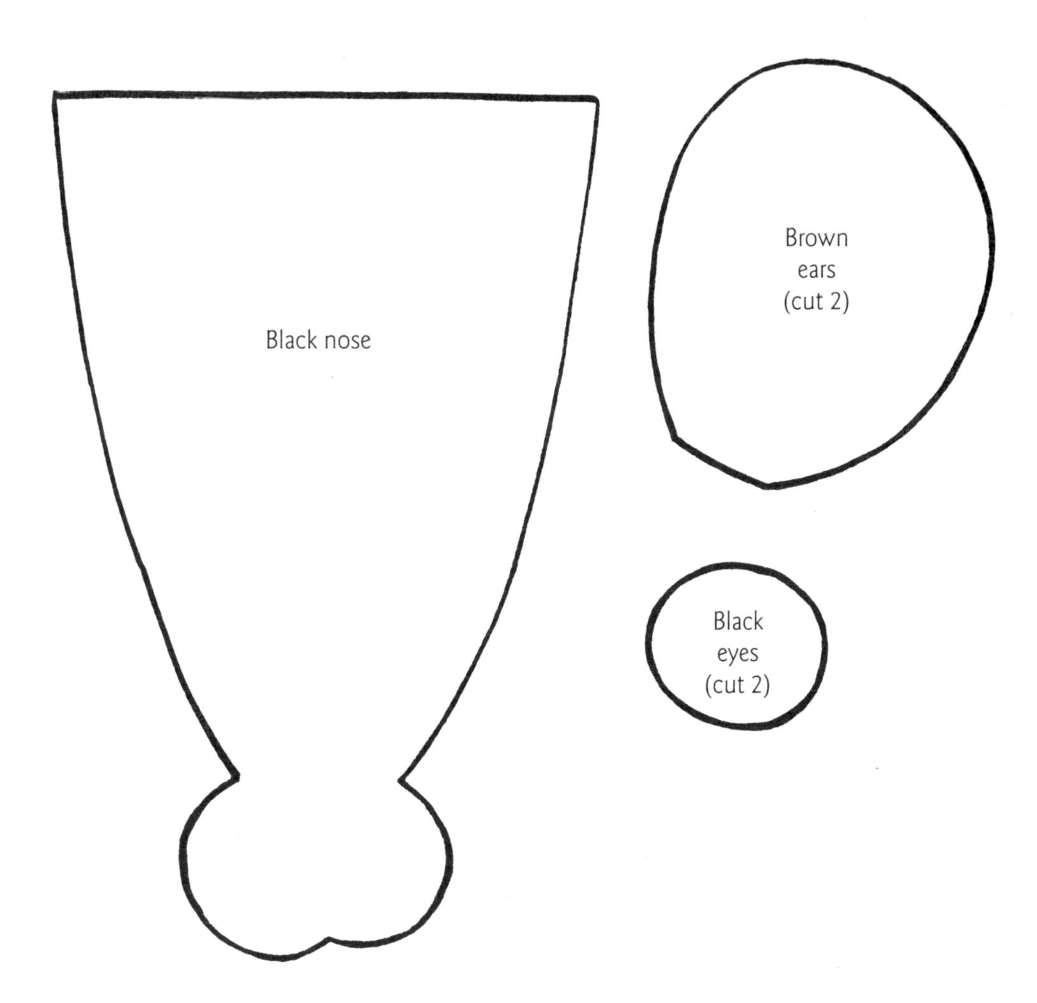

■ **3-33** *The patterns for the pocket opossum.*

Directions

1. Create an opossum head by coloring the bottom of the paper plate brown and the top of the paper plate pink. Fold the plate in half with the brown on the outside.
2. Use the patterns in Fig. 3-33 to trace and cut out two black circle eyes, two brown ears, and a black nose. To make an opossum face, look at Fig. 3-32 as a guide for gluing these pieces onto the plate.
3. The remainder of the opossum is made of pieces cut from the grocery bag. To create the opossum's body, use the ruler to measure and draw a rectangle $10\frac{1}{2}$ by 14 inches on the bag. Cut it out. Round off the corners with your scissors. Staple the paper plate head onto the body as shown in Fig. 3-34.
4. To make the mother opossum's pocket, use the ruler to measure and draw a rectangle $6\frac{3}{4}$ by $7\frac{1}{4}$ inches on the bag. Cut it out. Fold down one long edge of the pocket $\frac{1}{2}$ inch; this edge will be its top. (Folding down the top of the pocket makes it stronger.) Glue it onto the body so that the pocket opens at the top.

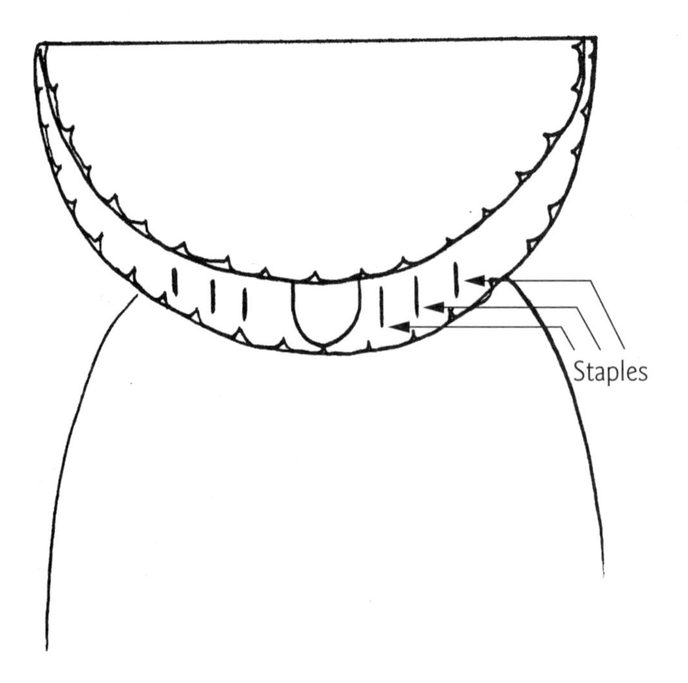

■ **3-34** *Staple the paper-plate head onto the body.*

5. The common opossum has five pink toes on each foot. To make the front legs, measure and draw two rectangles 2 by 6 3/4 inches on the grocery bag. Cut them out. Draw and cut five fingers on the end of each leg. Color them pink. Glue them on the body close to the head.
6. You can make the back paws from two ovals shaped like potatoes. Draw lines to show the five toes. Glue them on the front of the body near the bottom.
7. The opossum has a very interesting tail. Not only is it long, hairless, and rope-like, it also is *prehensile,* which means the opossum uses it to hold onto things like tree branches. Design an opossum tail from grocery-bag paper. It should be half as long as the body. Glue the tail to the front edge of the body so that it can be seen.
8. Use grocery-bag paper to make opossum-baby finger puppets. For each opossum puppet, measure and cut out a 3-inch paper square.
9. Glue each square into a cylinder shape that fits easily over your finger. Here's a hint to make gluing easy: Wrap the squares tightly around a pencil, and then remove the pencil before you glue.
10. Glue on two black eyes and a pink nose made with a paper punch. Glue on little round paper ears. Add 1 1/2-inch-long prehensile tails that look hairless and rope-like. After the glue dries, put the babies in the mother's pouch.

☞ *Closure*

Following directions; using numbers

Review your goals and give children opportunities to talk about what they have learned. Here are some finger-play counting rhyme activities for young children and a word search for older children.

Let the children use their fingers or make finger puppets to use as they say these rhymes.

Rhymes

Four raccoons

Looking out of a tree

One came out

Then there were three.

Five fireflies

Flying by the door

One came inside

Then there were four.

Three baby owls

Going "Whoooo, whooooo"

One flew away

Then there were two.

Four beavers

Gnawing on a tree

One got tired

Then there were three.

Word Search
Nocturnal Animals

Find the hidden words. Draw a circle around each nocturnal animal you can find. All of the words are going across. See if you can find these words:

- Owl
- Raccoon
- Beaver
- Bat
- Opossum
- Moth
- Firefly

```
a d b t o w l m j k r w a p l o e q r a c c o o n w
m w r y u o k m b e a v e r x v k o u m b n h j t
r b a t g y u i o p a z x s w e r v g o p o s s u m
m o t h v x t h m k l o i u e w q a p l m n y c v d
x g e h f i r e f l y m i j h t c x w q a z p i y r w
```

chapter 4

Classroom Camping Adventure

Science goal

To help children learn about night and traditional nighttime experiences by camping out in the classroom

Planning

Teacher planning is very important in this chapter. The teacher needs to carefully plan and then let the children and parents help create the setting.

Get a plain piece of drawing paper. Read this chapter through in its entirety. As you read, draw the setting on paper as it might appear in your own classroom. Make a list of the specific things you want to use. Send a letter home to parents explaining the camp out and ask them to send in or bring specific items by a certain date.

The following are ideas for setting up the environment for camping out in the classroom:

1. *Make a tape of night noises.* Prepare an easy tape of night sounds by taking a tape recorder out at night and letting it run for an hour or so. You should get sounds such as car noises, crickets, owls, dogs barking, and people talking. One of the children might like to prepare this tape for the class to use.
2. *Create a starlit sky.* Get several strings of small white holiday lights (the kind used on Christmas trees). Attach them to the ceiling of the room. One easy way to do this is to first attach fishing line from various parts of the ceiling. At the end of the fishing line, attach a clothespin. Now you can clip the strings of lights with the clothespins. The ceiling will become a very pretty "starlit sky."
3. *Make a balloon moon.* If you want to hang a moon, you can blow up a large white balloon and use a clothespin to hold it as you did with the lights. When the lights in the room are off, you can shine a bright light, like that of an overhead projector, onto the moon and create different phases.
4. *Darken the windows.* If you have windows in the room, you can cover them with dark blue or black craft paper. Before you put the paper up, you can either punch small holes in the paper with a sharp pencil point or other suitable object, or you may want to let the children do it. These holes can be stars—the daylight will shine through. You can make stars at random or make constellations.
5. *Add bushes and trees.* Let the children create small bushes (about 1 foot tall) and large trees (perhaps about 5 feet tall) out of craft paper. Attach them to the wall. For the trees, cut trunks and limbs from brown paper. Let the children sponge-paint various colors of brown and gray onto this brown paper to give it the effect of bark. Leaves can be made from construction paper (green leaves or colored leaves for fall) and glued onto the limbs of the trees. If you can get real or artificial trees in pots, you can add these to the setting also. For the small bushes, you can cut large circles of green craft paper and let the children sponge-paint various shades of green onto the basic green. The children may enjoy drawing and coloring pictures of nocturnal animals and placing them on and around the trees.

6. *You'll need some tents.* In your note to parents, ask them to send tents if possible. You can set up several of these in the room. The children will enjoy using them during the day or pretend night. If you can't get real tents, you can use sheets or blankets. Hang a rope from two points (about 5 feet apart). The rope should be at least 4 feet from the floor. Drape the sheet over the rope so that the middle of the sheet is on the rope. Secure the sides of the sheet on the floor with bricks or large rocks.
7. *Get flashlights.* Have flashlights available. You can ask for these to be sent in your letter.
8. *Make nocturnal animals.* Help the children make the models of nocturnal animals described in Chapter 3. These animals can be placed around the room and can be used in all kinds of dramatic play.
9. *Prepare for camping snacks.* Gather the ingredients for the cooking activities described later in this chapter.
10. *Talk about camping clothes.* Think about what types of clothes you may want the children to wear on the day or days of their camp out. Comfortable, hiking-type clothing would be fun. You may want to let the children bring their bedroom shoes to wear for part of the time.
11. *What else do you need?* Decide on other items you want the children to bring. Some suggestions are sleeping bags, pillows, blankets, stuffed animals that they like to sleep with, bedtime story books, and alarm clocks.

Let's create a classroom campfire

Following directions; manipulating materials; creating a model

It's fun to sit around a classroom "campfire" that's almost as good as the real thing!

What you need

- ☐ Black bulletin board paper
- ☐ A battery-powered camping lantern or three flashlights
- ☐ Red cellophane
- ☐ Six real logs or large sticks

Directions

1. Cut a large circle, about 24 inches across, from the black bulletin board paper. Lay this circle on the floor where you are building your campfire.
2. Put the camping lantern or flashlights in the center of this paper circle. Drape a sheet of red cellophane over the lights to cover them loosely. Wad up several more pieces of cellophane and black paper and add those to the fire to look like red and black coals.
3. Place the logs carefully over the lantern so that when it is turned on, you can see the red glow through the logs.
4. You can sing songs and tell stories around your classroom campfire. You can also pretend to cook by putting cooked hot dogs on sticks and holding them over the "fire" before eating them.

Nighttime Note *With black-and-white photography, document the work of creating the classroom night scene. Encourage the children to help you think of a caption to write under each photograph, describing the action and naming the people in each picture. Use them to create a picture wall for everyone to enjoy. Send the individual pictures home with the children when the nighttime study ends.*

Let's make some camping necessities

Campers need a lot of things to take with them into the woods. Our classroom campers can make their own binoculars, backpacks, visors, nighttime goggles, and day-and-night camping plates. Campers can also build their own "classroom campfire."

Let's create a paper-bag backpack

Measuring; manipulating materials; following directions
Create your own backpack from a grocery bag (Fig. 4-1). Add cloth straps and fill your pack with handmade camping supplies.

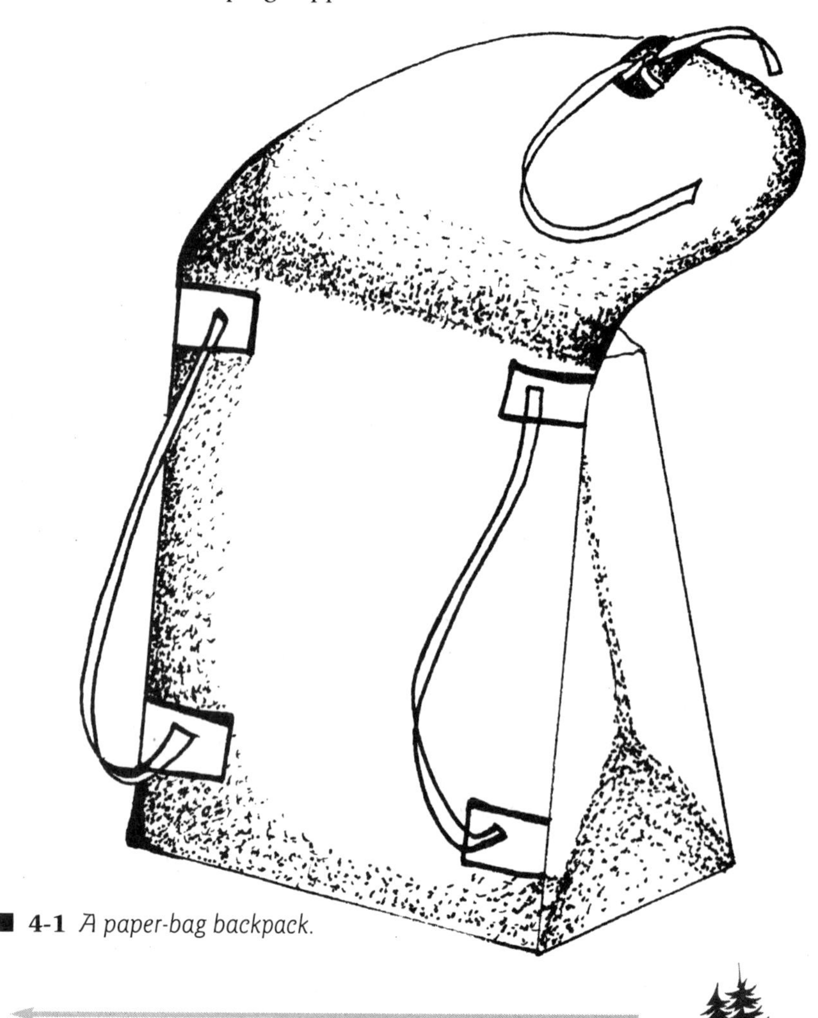

■ **4-1** *A paper-bag backpack.*

What you need

- ☐ Heavy paper grocery bag
- ☐ Ruler
- ☐ Scissors
- ☐ Markers
- ☐ Paper punch
- ☐ Clear packing tape
- ☐ 24-inch ribbon or yarn
- ☐ Cloth to be cut into two backpack straps, $2^1/_2$ by 24 inches each

Directions

1. First create a flap for the backpack. Open the bag. Beginning at the top of the bag, use a ruler to measure and draw two lines each 7 inches long down the back folds of the bag (Fig. 4-2).
2. Cut on these lines. Then cut a rounded edge on the flap (Fig. 4-3).

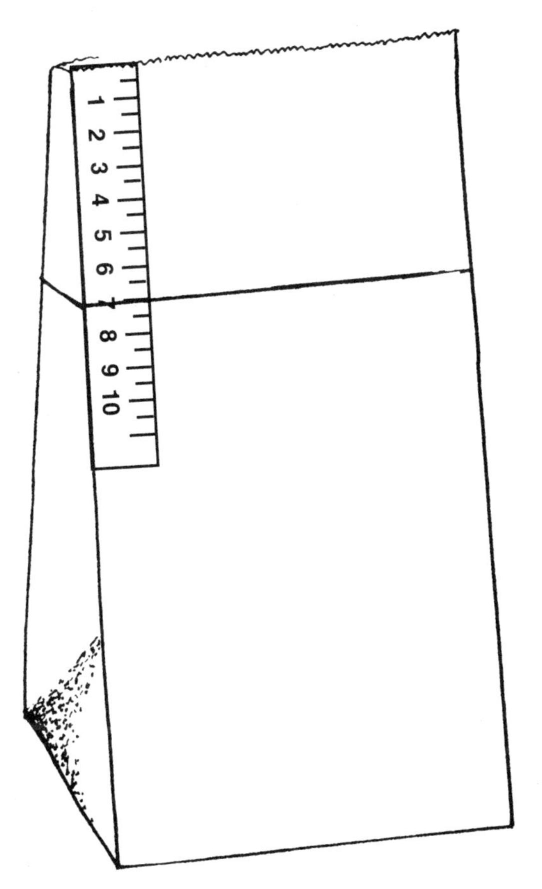

■ **4-2** *Measure two lines 7 inches long down the back folds of the paper bag.*

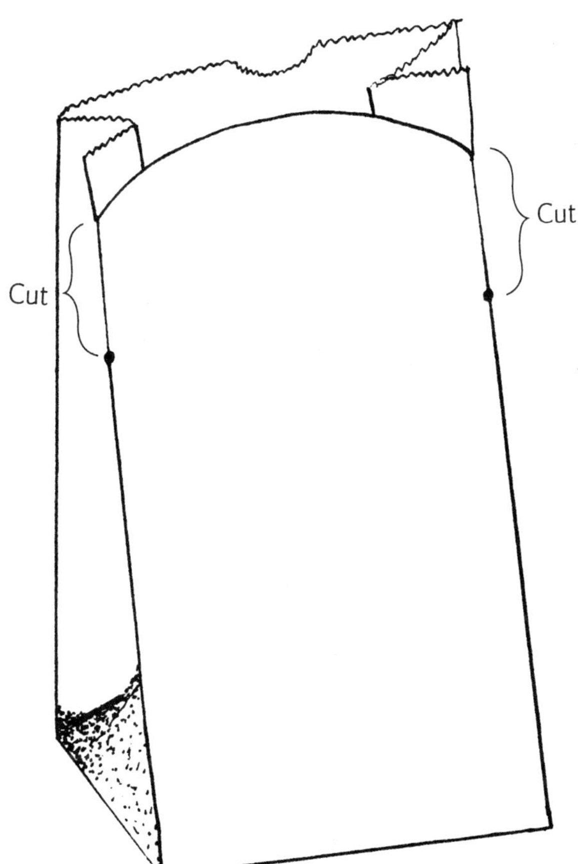

■ **4-3** *Cut a round edge on the flap.*

3. Next, the rest of the bag must be trimmed in a straight line all the way around. To do this, use the ruler to measure 7 inches from the top of the bag. Draw dots all around the bag 7 inches from the top. Then use the ruler and a marker to connect the dots in a straight line around the bag. Cut on the line (Fig. 4-4).

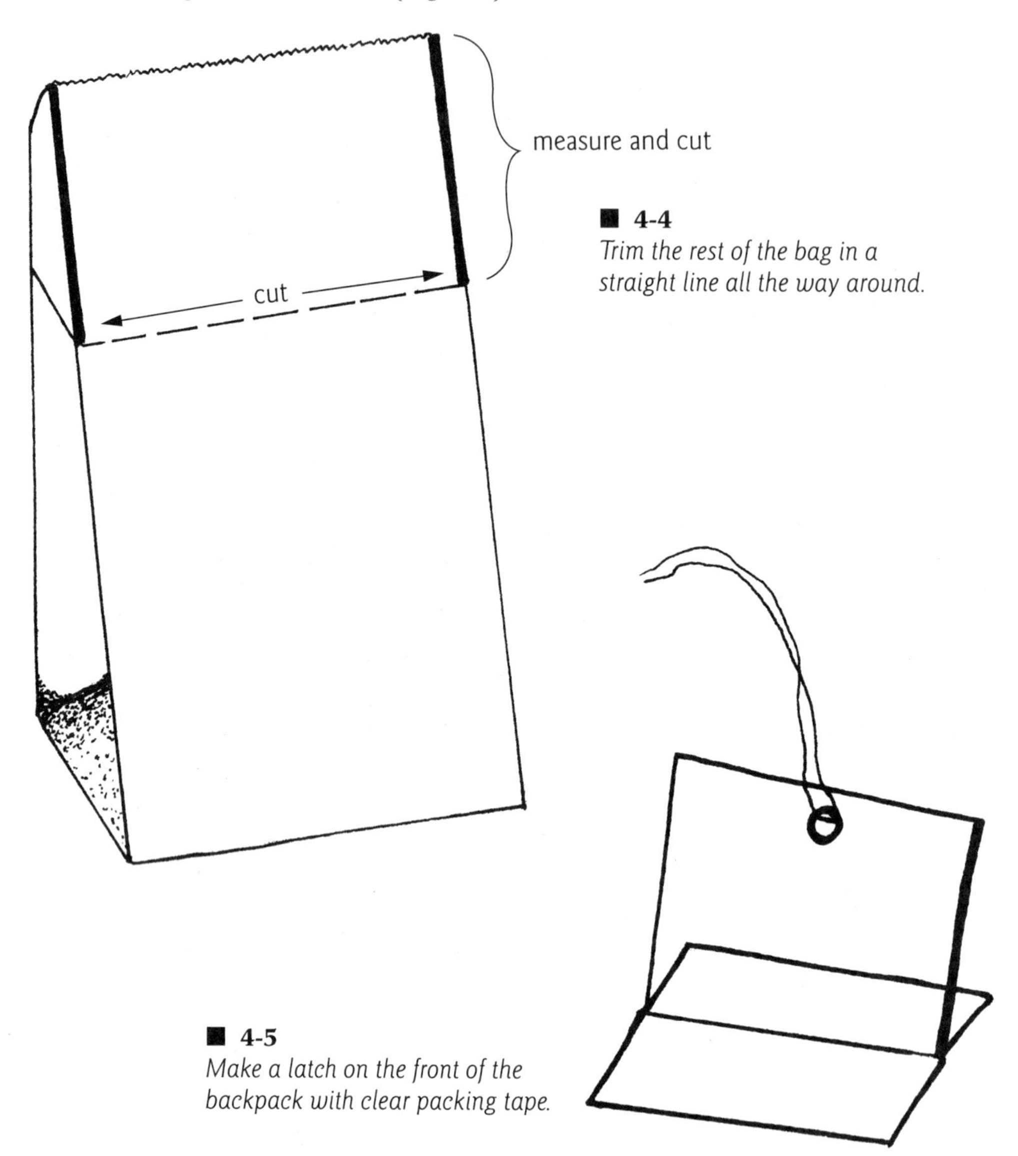

■ **4-4**
Trim the rest of the bag in a straight line all the way around.

■ **4-5**
Make a latch on the front of the backpack with clear packing tape.

4. You can make a latch on the front of the backpack with a 4-inch piece of clear packing tape. Fold the tape onto itself, leaving the two ends spread open (Fig. 4-5). Stick this tape to the front of the backpack. Punch a hole in the middle of the latch, and tie on a 12-inch piece of ribbon or yarn.
5. Complete the latch by putting packing tape on the front edge of the flap, top and bottom. Punch a hole in it and tie on another 12-inch piece of ribbon or yarn.
6. Tape the cloth straps to the back of the backpack (see Fig. 4-1).

Let's create paper-plate nighttime glasses

Becoming more aware of the differences between daytime and nighttime; following directions; manipulating materials; creating a model

Night-vision glasses help soldiers and pilots see better in the dark. These nighttime glasses make everything look like night! The frames are made of a paper plate, and the lenses are colored cellophane.

What you need

- ☐ 9-inch paper plate
- ☐ Scissors
- ☐ Markers
- ☐ White glue
- ☐ Cellophane tape
- ☐ Dark-blue cellophane or blue plastic wrap

Directions

1. Use the patterns in Fig. 4-6 (shown on the next page) to trace the frame and two earpieces onto a paper plate (Fig. 4-7). Decorate them with markers and cut them out.

■ **4-7** *Trace the frame and earpieces onto a paper plate.*

■ **4-6** *The patterns for the nighttime glasses.*

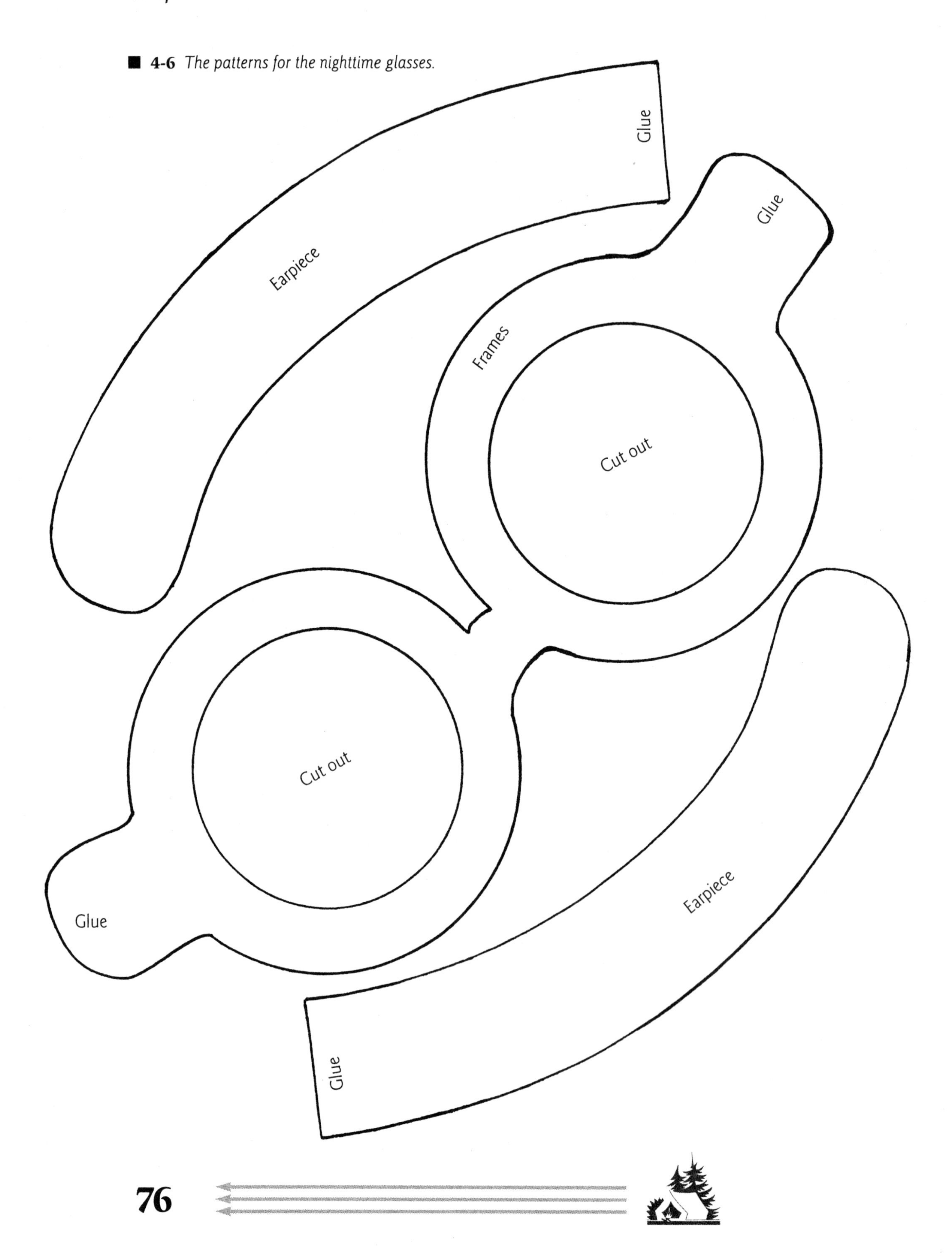

2. Glue the two earpieces on the frames. Let the glue dry.
3. Lay the glasses face up on a sheet of blue cellophane or plastic wrap. Tape the frames to the cellophane as shown in Fig. 4-8. Trim the glasses, leaving a small border of cellophane around the frame edges (Fig. 4-9).
4. Put your glasses on to make daytime turn into night. Remember, at night you can't see very well, so be careful not to bump into things!

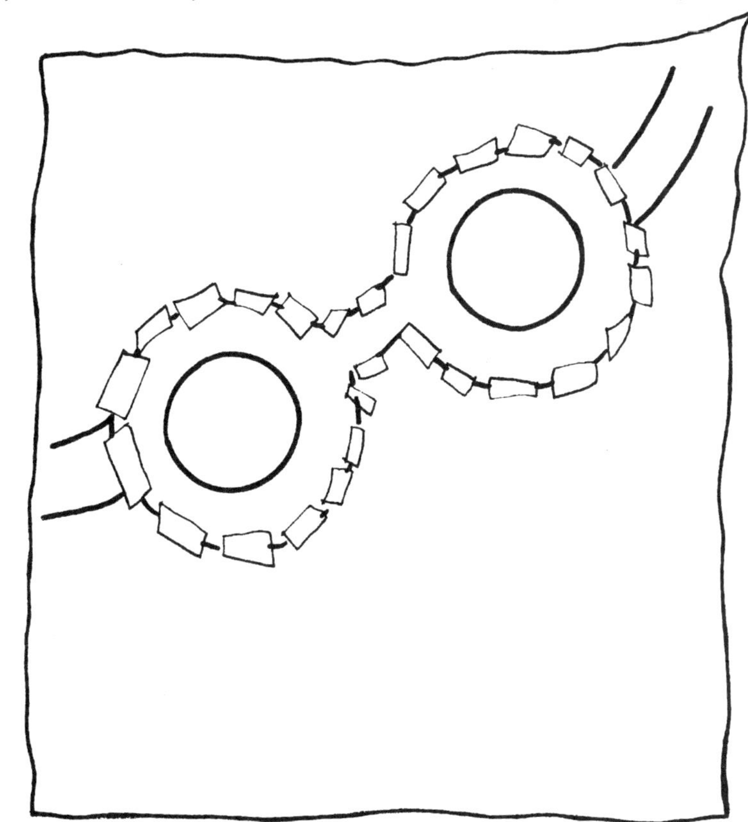

■ **4-8**
Tape the frames to cellophane.

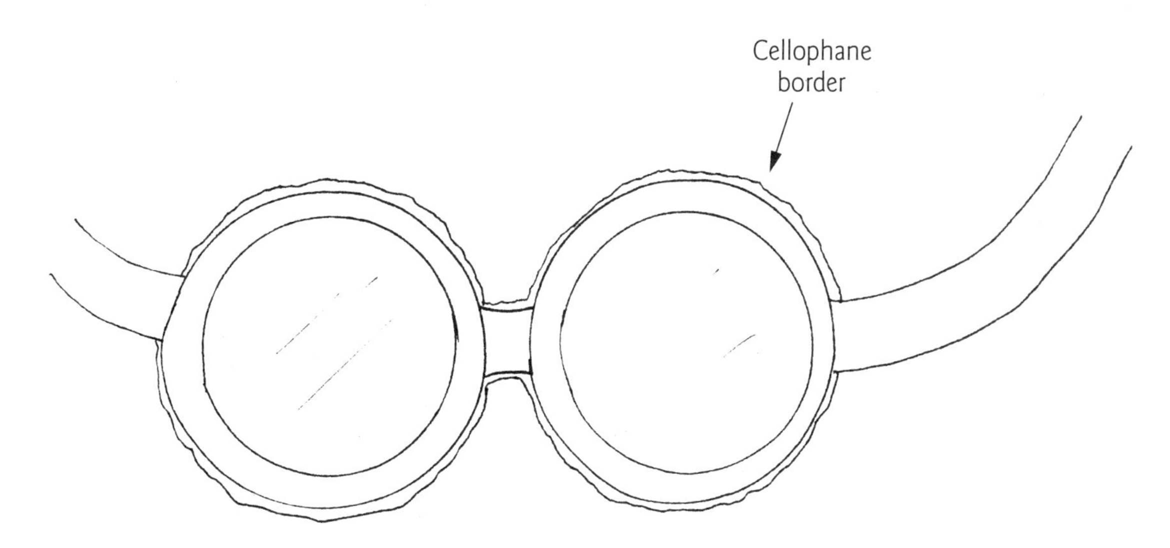

■ **4-9** *Trim the glasses, leaving a small border of cellophane around the edges.*

Let's create a paper-plate visor

Finding and recording information; manipulating materials; following directions

A visor is a good thing to wear when you hike and camp. It keeps your hair out of your eyes while you're watching the trail. Decorate your visor with pictures of nocturnal animals (Fig. 4-10).

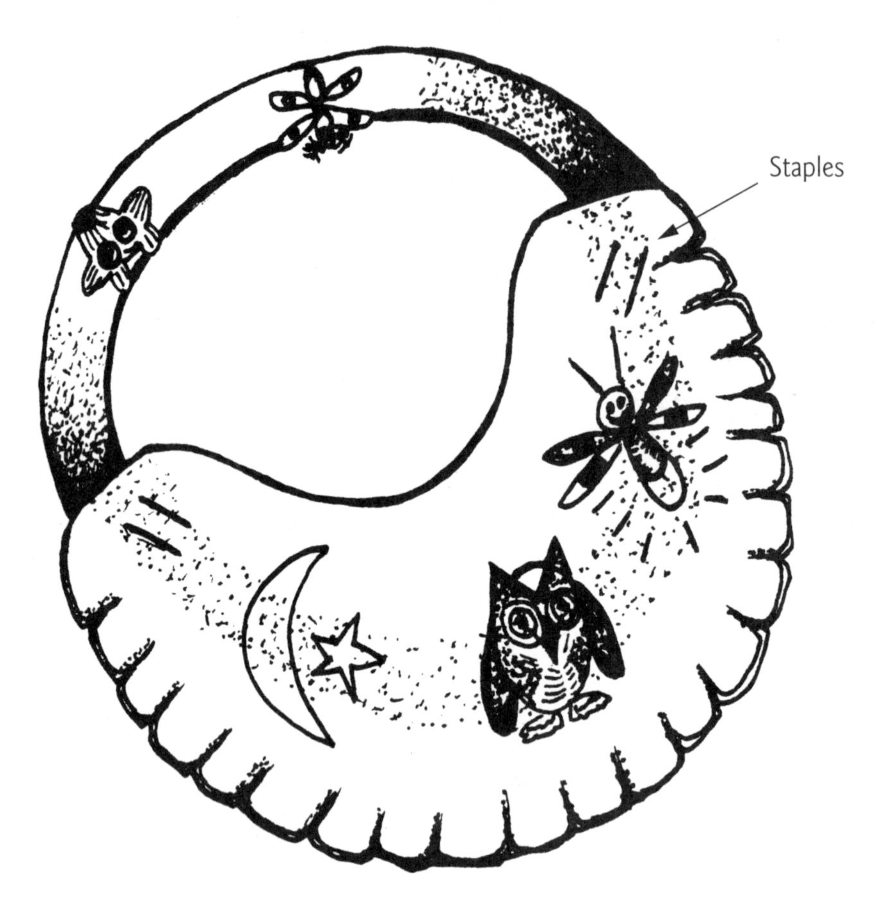

■ **4-10** *A paper-plate visor.*

What you need

- ☐ Some pictures of nocturnal animals
- ☐ 9-inch paper plate
- ☐ File folder
- ☐ Scissors
- ☐ Markers
- ☐ Stapler

Directions

1. Use the visor pattern in Fig. 4-11 to trace and cut out a visor from a 9-inch paper plate.

4-11 *The pattern for the paper-plate visor.*

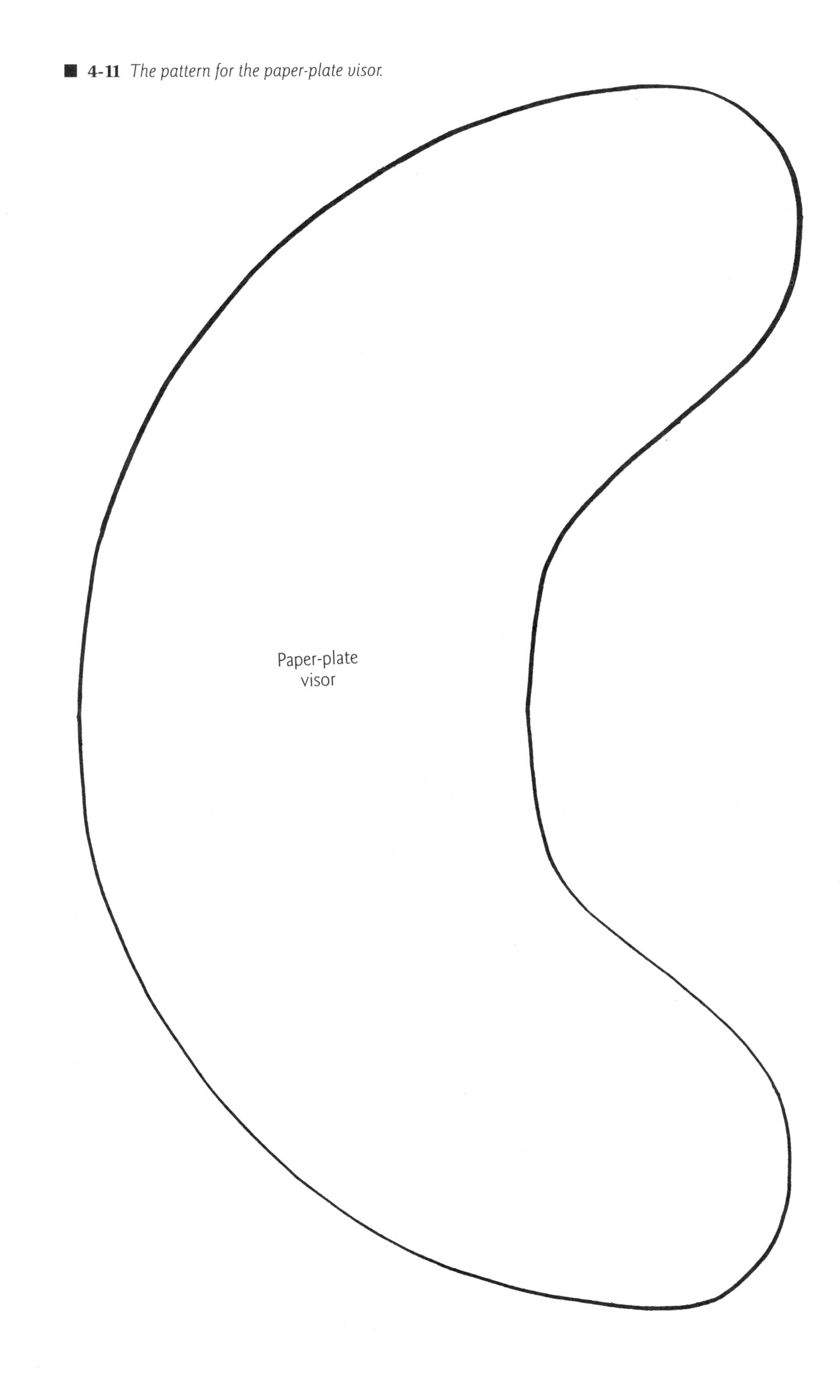

■ **4-12** *The pattern for the visor strap.*

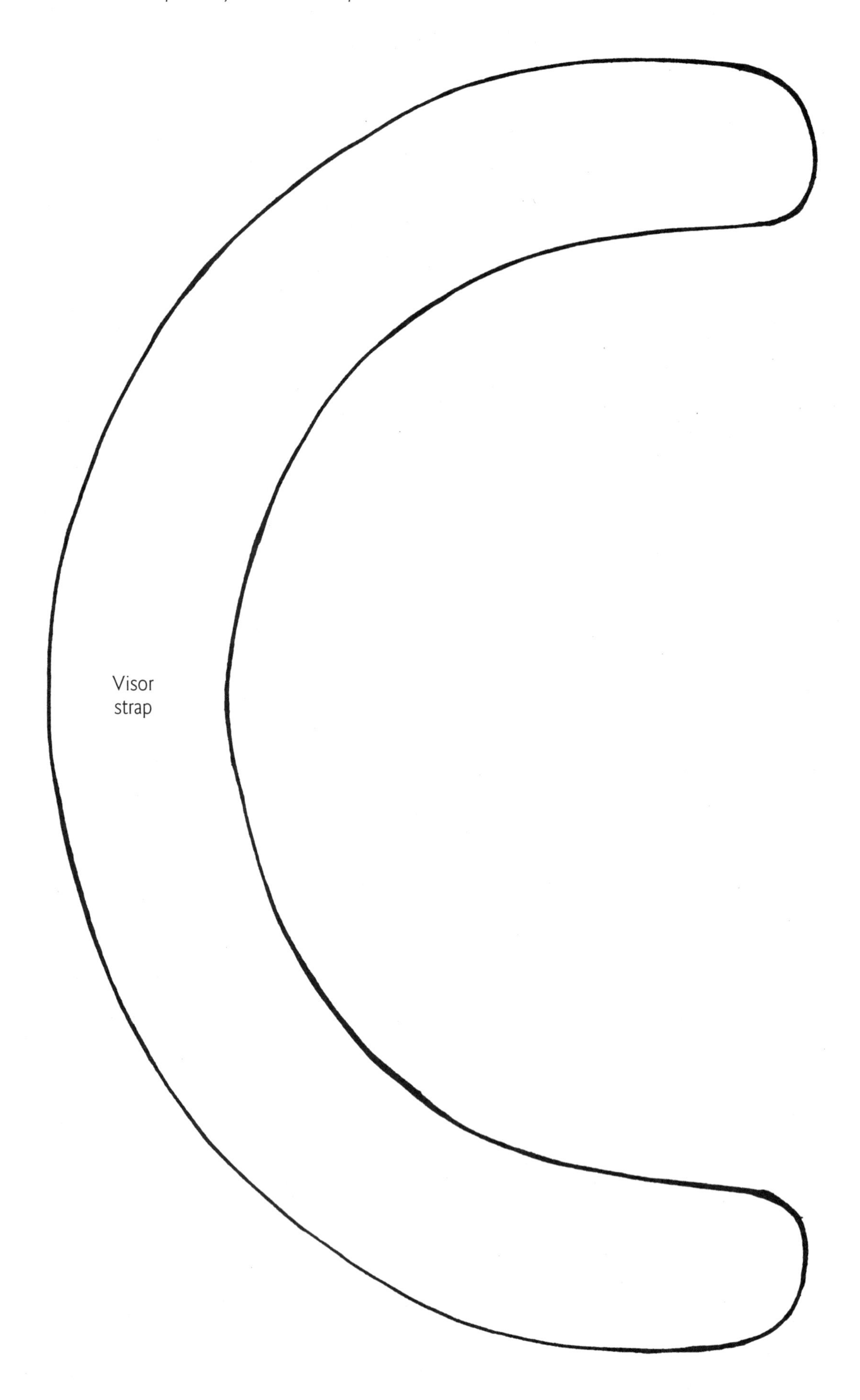

2. In your library or media center find some pictures of nocturnal animals. Draw some of these animals on your visor.
3. Use the visor strap pattern in Fig. 4-12 to trace and cut out a strap from a file folder. Decorate it with more nocturnal animals. Staple the visor strap to the visor as shown in Fig. 4-10. Adjust it to fit your head size before stapling it.

Let's create cylinder binoculars

Following directions; creating a model; manipulating materials; measuring

Real binoculars help campers see things that are far away. Although our model binoculars can't do that, they're still fun to make and take along on our classroom camp out.

What you need

- ☐ Binoculars
- ☐ Three toilet-paper roll cylinders
- ☐ Ruler
- ☐ Construction paper
- ☐ Scissors
- ☐ Markers
- ☐ White glue
- ☐ Two large paper clips
- ☐ Paper punch
- ☐ Ribbon or yarn

Directions

1. Examine real binoculars so that you can make your model binoculars look as real as possible.
2. Use the ruler to draw two rectangles $4\frac{1}{2}$ by 6 inches on construction paper. Cut them out. Spread glue on one rectangle. Wrap it around one cylinder, covering the cylinder completely. Repeat this step with the second rectangle and cylinder.
3. Cut the third cylinder into narrow circles and use them to help glue the binoculars together. First squeeze the remaining cylinder until it is flat. Then cut a $\frac{1}{2}$-inch piece from one end.
4. Flatten the piece, spread glue on both sides of it, and sandwich it between the two covered cylinders (Fig. 4-13).

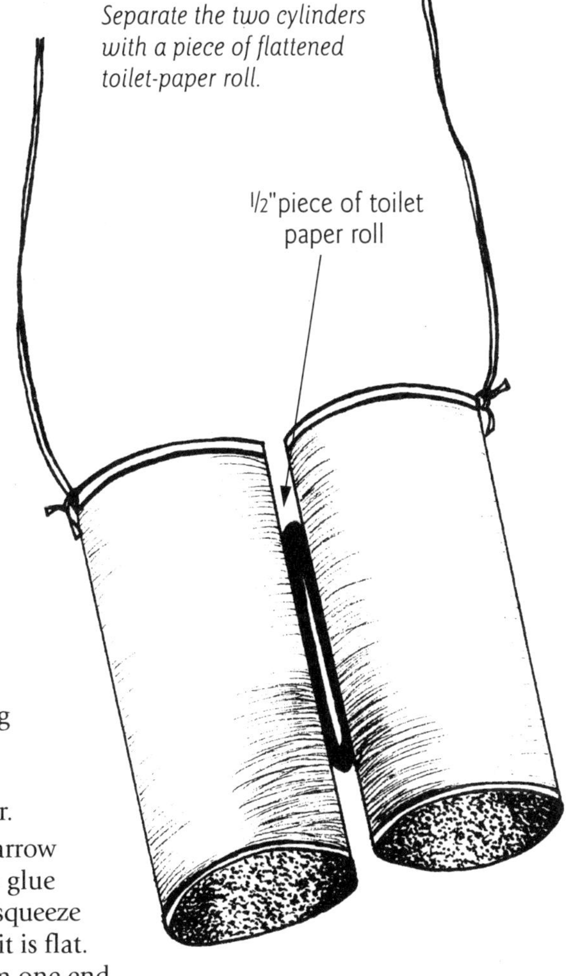

■ **4-13**
Separate the two cylinders with a piece of flattened toilet-paper roll.

5. Hold the two cylinders together at both ends with a large paper clip while the glue dries for 10 minutes.
6. Remove the paper clips. Color your binoculars with markers.
7. Punch two holes near the top. Tie on a length of ribbon or yarn that will fit over your head to keep the binoculars around your neck.

Let's create a day-and-night camping plate

Applying knowledge; following directions; creating a model; manipulating materials; using a ruler

By coloring and cutting two plain paper plates, you can create a camping scene that turns from day into night (Fig. 4-14).

■ **4-14** *A day-and-night camping plate.*

What you need

- ☐ Two 9-inch white paper plates
- ☐ Ruler
- ☐ Scissors
- ☐ Crayons
- ☐ Lightweight cardboard
- ☐ Green tissue paper
- ☐ White glue
- ☐ Stapler
- ☐ Stick-on stars

Directions

1. Draw a light line with a pencil halfway across the middle of each paper plate. Cut each plate on this line, from the outside edge to the center (Fig. 4-15).

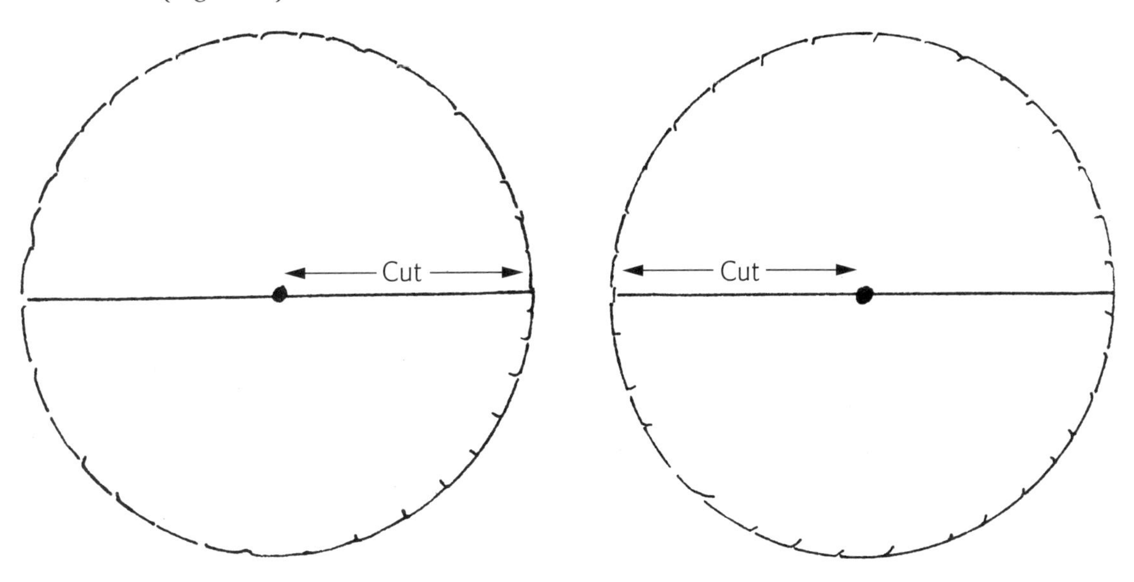

■ **4-15** *Draw a line halfway across the center of each plate. Cut from the edge to the center.*

2. Put the two plates together by sliding one cut side into the other. Notice how you can rotate one plate over the other (Fig. 4-16). Separate the plates.

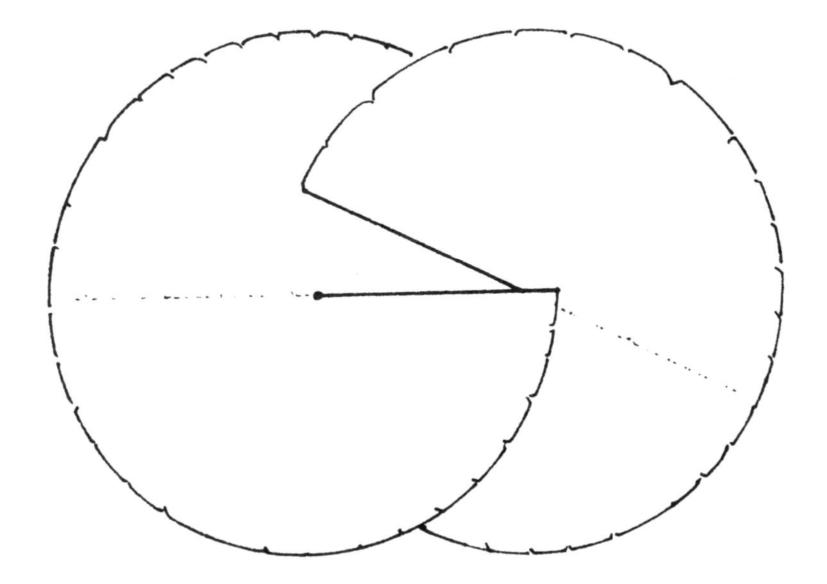

■ **4-16** *Put the plates together by sliding one cut side into the other.*

■ **4-17** *The patterns for the day-and-night camping plate.*

3. Turn one plate into a daytime campsite by coloring the top half of the plate blue for sky. Use the tent pattern in Fig. 4-17 to draw a tent onto the bottom half of the plate. Color green grass around the tent as shown in Fig. 4-18. Cut a slit in the tent so that the camper can slide in.

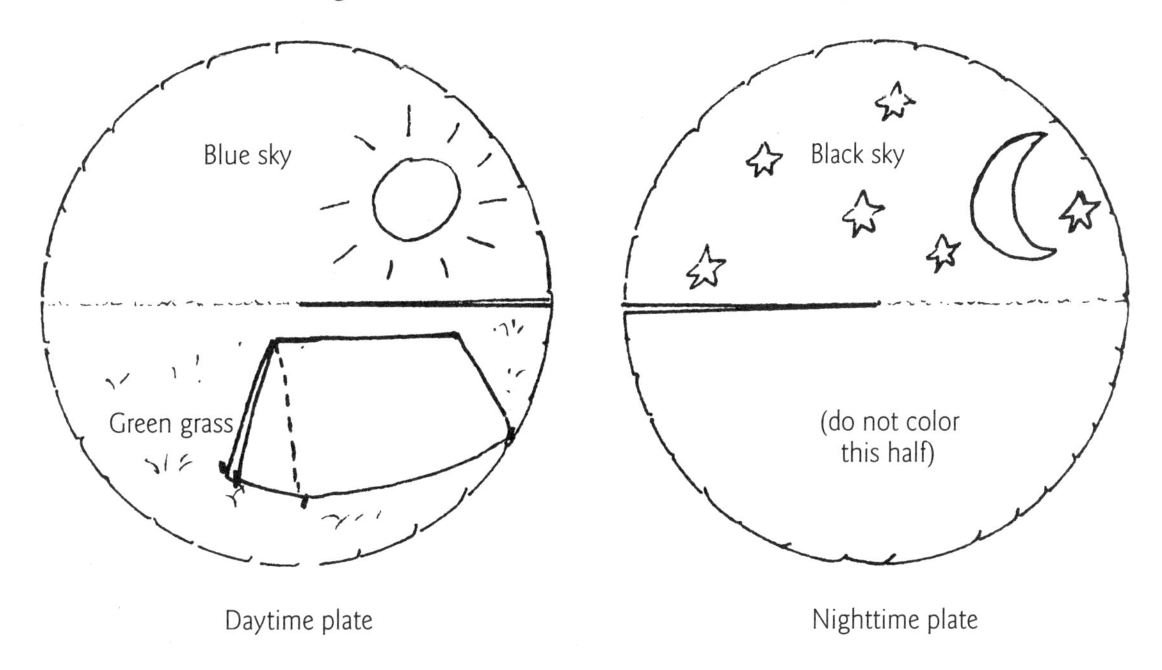

■ **4-18** *Color the daytime plate and half of the nighttime plate.*

4. Turn the other plate into a nighttime plate by coloring only the top half of the plate with a moon and a dark sky. Add stick-on stars (Fig. 4-18).
5. Use the patterns in Fig. 4-17 to trace the tree, camper, and campfire onto lightweight cardboard. Cut these out and color them. We colored one side of the camper so that he or she was asleep with eyes closed. We colored the other side of the camper with eyes open.
6. Glue the fire onto the plate. Staple the tree onto the green half of the plate. Let it stick up into the sky as shown in Fig. 4-19. We added little bits of green tissue paper to our tree as leaves.

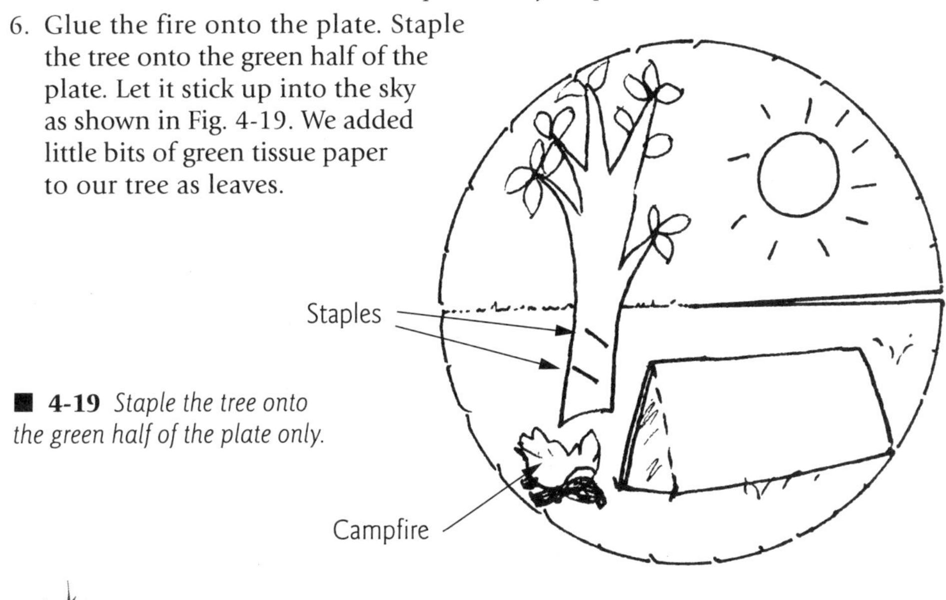

■ **4-19** *Staple the tree onto the green half of the plate only.*

7. Put your finished plates together by sliding one over the other again. Rotate the plates to turn the campsite into a daytime scene or a nighttime scene (Fig. 4-20).
8. Your camper can be awake at the campsite during the day. At night you can tuck him or her into the tent with eyes closed!

■ **4-20**
The finished daytime and nighttime plates.

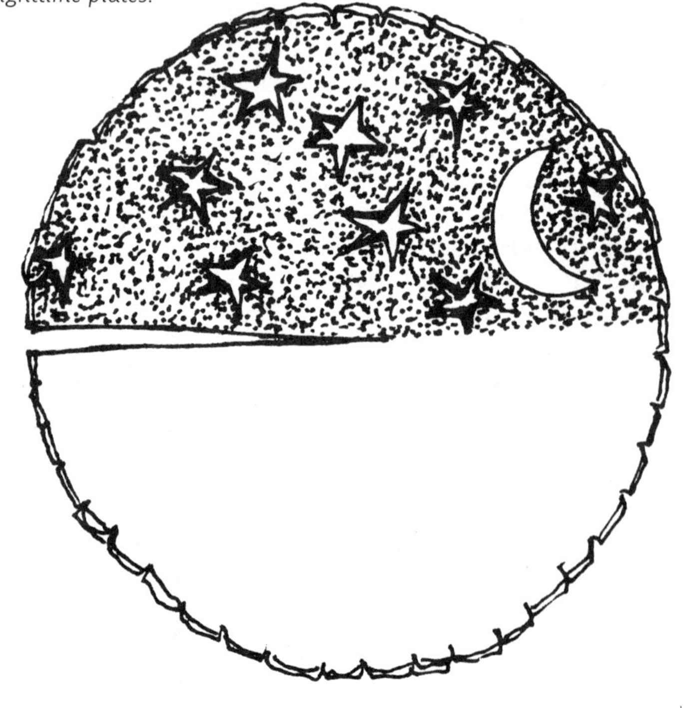

Snacks for camping out

After setting up a tent, building a campfire, and making their own camping gear, our classroom campers will be hungry. Here are some campfire-perfect recipes that are easy to make in the classroom.

Let's create some campfire recipes

Observing changes in foods; following directions; manipulating materials; measuring

It's fun to sit around a classroom campfire and snack on real campfire foods that you've prepared yourself. Here are some easy campfire food ideas that can be prepared in the classroom.

Baked potatoes

1. Wash and dry one small potato per person. Wrap each one in aluminum foil.
2. Put the potatoes in a covered electric crockpot and cook on the highest setting until they are soft. It is not necessary to add water to the crockpot.
3. Hide the cooked potatoes in the classroom campfire and pretend that they cooked among the black paper "coals."

Hot dogs

1. Put hot dogs in a covered electric crockpot with $1/4$ cup of water.
2. Cook them on the highest setting until they are hot.
3. Insert a craft stick in each hot dog. Hold it over the classroom campfire and pretend to cook it. Put it in a bun.

Hot chocolate

1. Use the electric crockpot to make hot chocolate with milk and instant chocolate mix or syrup.
2. When the mixture is warm, pour it into serving cups with a few miniature marshmallows added. Drink it around the classroom campfire.

French toast with powdered sugar

1. Melt 2 tablespoons of margarine in an electric skillet.
2. Mix two eggs and $1/2$ cup milk in a bowl.
3. Dip bread slices quickly into the egg-milk mixture, and fry each slice on the skillet until it's golden brown on each side.
4. Cut the slices into strips and sift on a little powdered sugar. Serve on paper saucers.
5. Each classroom camper can sit around the campfire and enjoy a real camper's breakfast.

Classroom s'mores

1. Spread marshmallow creme on a graham cracker square.
2. Sprinkle the marshmallow with miniature chocolate chips.
3. Put another graham cracker square on top.

Campers' great granola

1. Set out individual bowls of raisins, granola cereal, nuts, and coconut.
2. Put a spoon in each bowl. Each classroom camper should have a plastic zipper-seal sandwich bag.
3. The campers can mix their own great granola by spooning some of each ingredient into their own bags.
4. The teacher may want to add a few miniature chocolate chips to each bag before "zipping" it shut.
5. The campers can shake their bags to mix the granola.

Campfire songs and stories

Children love sitting around a campfire, singing songs, and telling stories. Get some tapes of lullabies to play as the children relax. Some original stories and a song your children will enjoy follow.

The Five Friendly Stars

This story can be presented with star and moon puppets as a puppet show (Figs. 4-21 and 4-22).

Once upon a time there was a little boy who was sick for a long time. He had to stay in bed. During the day he would read and play games in bed, but when it got dark, he liked to look out his window at the stars.

Well, at this same time, there were five stars up in the sky who were all good friends with each other. They were sad. They wanted to be special and have a name like the constellations have—like the "Big Dipper."

The Moon, who knew that the little boy loved to look at the stars, came to the five stars and asked them to help cheer the little boy up by doing something special. The stars were excited. They had a meeting. They all held hands. What could they do? How could they get the little boy's attention?

The next night, when the little boy was looking up in the sky, the five friendly stars decided to make shapes in the sky, in hopes that the little boy would see them. The moon had told them that the little boy really loved trains, so the five friendly stars lined up like a train. The little boy looked up into the sky and saw the star train. He was so excited! He smiled and began to feel much better. In a few days he was well, but every night he still looked for his special star train.

The five friendly stars were happy because they had done something special for the little boy. The moon said that these stars could have the name Star Train forever, and they all lived happily ever after.

The Camping Adventure

One day ____________ and ____________ decided to go on a camping adventure. They were very ____________ about their trip. "Let's take some ____________ to eat," said ____________ . "Yes, and let's also take ____________ , ____________ , and ____________ ," said ____________ .

Soon they were ready. They walked and walked. It seemed they walked for ______ ____________ miles. Along the way they saw____________ , ____________ , and ____________ . They picked some ____________ and ____________ to take along with them.

Soon they came to the perfect camping spot. The ground was covered with ______ ____________ and tall ____________ grew all around. "Let's pitch our tent here," said ____________ . "I'll get wood for a fire," said ____________ .

When their campsite was ready, they began to prepare supper. They ate their favorite foods, ____________ , ____________ , and ____________ . They had brought lots of ____________ to drink.

As darkness settled around them, they listened quietly to ____________ . "I guess it's time to go to sleep," said ____________ . "I'm sleepy, too," said ____________ .

No sooner had they settled into their sleeping bags and closed their eyes when they heard a soft ____________ . "Did you hear that?" whispered ____________ . "Yes, what do you think it is?" ____________ whispered back. Then they heard ______ ____________ again. It was louder this time and closer. "I have to see what that is," said ____________ .

"It might be a big ____________ !" said ____________ . They peeked around the edge of the tent. What do you think they saw?

To their surprise, it was their mom!

"I came to see if you two needed another blanket, and I brought the flashlights you forgot."

"Gee, thanks, Mom," said ____________ and ____________ . "We're glad we decided to camp out in our own backyard."

■ **4-21** *The pattern for the moon stick puppet.*

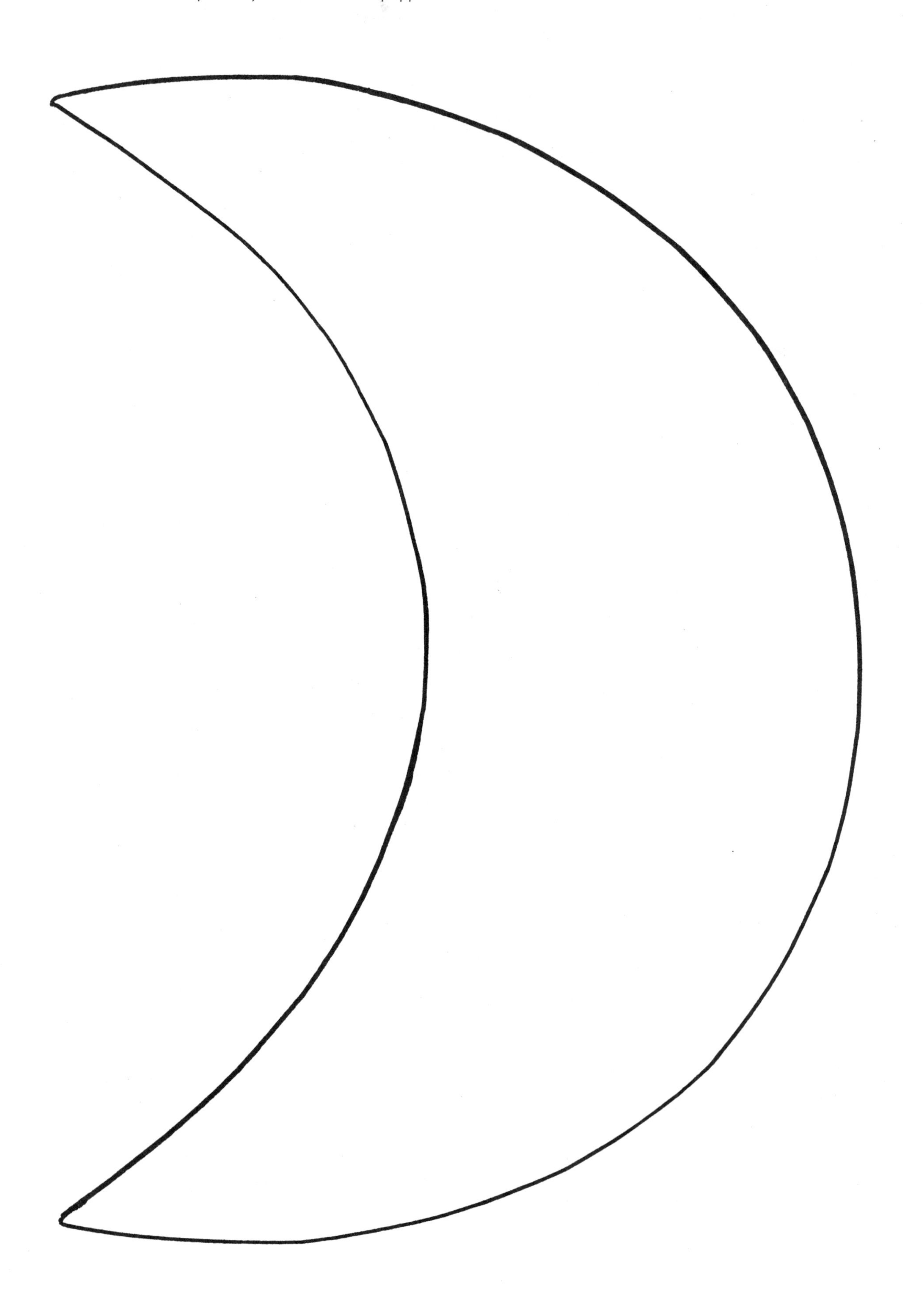

■ **4-22** *The patterns for the star stick puppets.*

Let's create star and moon stick puppets

Following directions; manipulating materials; creating a model
Use these stick puppets in a class play.

What you need

- ☐ Poster paper or a file folder
- ☐ Scissors
- ☐ Crayons
- ☐ Glitter
- ☐ White glue
- ☐ Masking tape
- ☐ Large flat stick (paint stirring stick, large craft stick, or ruler)

Directions

1. Use the patterns in Fig. 4-21 and 4-22 (shown on the preceding pages) to trace a moon and at least three stars.
2. Cut them out of poster paper or file folders.
3. Color them with crayons. Spread on a little bit of glue and sift on some glitter.
4. Use masking tape to fasten each star and moon to a stick such as a large craft stick, a paint stirring stick from a hardware store, or a ruler.
5. Now you're ready to take part in a play about a starry night.

"At Night It Is Dark"

This song is sung to the tune of "The Farmer in the Dell." One child is the Camper. He or she stands in the middle of the play area. The rest of the children join hands to form a ring around him or her while they walk slowly in one direction. Everyone sings.

At night, it is dark.

At night, it is dark.

The sun has gone, the moon has come.

At night, it is dark.

(The Camper stands in the middle of the circle.)

I can see the stars.

I can see the stars.

They twinkle softly in the sky.

I can see the stars.

(The Camper chooses a "star" to join him or her in the circle.)

I can see shadows.

I can see shadows.

Some are tall and some are small.

I can see shadows.

(The star chooses a shadow to join them in the circle.)

Owls are awake.
Owls are awake.
They are busy in the night.
Owls are awake.
(The shadow chooses an owl to join them in the circle.)
Raccoons are awake.
Raccoons are awake.
They are looking for some food.
Raccoons are awake.
(The owl chooses a raccoon to join them in the circle.)
Bats are awake.
Bats are awake.
They are flying here and there.
Bats are awake.
(The raccoon chooses a bat to join them in the circle.)
Fireflies are awake.
Fireflies are awake.
Their little lights go on and off.
Fireflies are awake.
(The bat chooses a firefly to join them in the circle.)
Beavers are awake.
Beavers are awake.
They're gnawing, gnawing on some trees.
Beavers are awake.
(The firefly chooses a beaver to join them in the circle.)
Opossums are awake.
Opossums are awake.
They're looking around for food to eat.
Opossums are awake.
(The beaver chooses an opossum to join them in the circle.)
I'm getting sleepy.
I'm getting sleepy.
I don't need to be awake.
I'm getting sleepy.
(The Camper pretends to fall asleep while the other characters rejoin their friends in the circle. Everyone sings the last verse together.)
Goodnight, moon and stars.
Goodnight, moon and stars.
I'll see you tomorrow night.
Goodnight, moon and stars.

Night noises

If children were actually in the woods, they would hear many night noises. The following activities can help them learn about night noises.

Encourage the children to take a tape recorder outside at night and to record some night noises. They should try to get sounds made by nature, if possible. Can they identify what is making the noises? Could it be a cricket or a tiny tree frog? They can bring the tape back to class and play it for their classmates. The children can draw pictures of the things that might be making the night noises.

Nighttime animals such as crickets and frogs use vibration to make noises and to communicate with each other. We can use vibration to try to imitate some night noises. One way to do this is to use a comb and waxed paper. Each child can fold a sheet of waxed paper in half and place it lightly over a regular hair comb. They should then place their lips very lightly on the waxed paper-covered comb, and make a long *uuuuhhh* sound with their voices. They should feel the paper vibrating and tickling their lips. Encourage the children to experiment with the sounds they can make. Can they sound like a cricket or another night noise maker?

Frogs make night noises by puffing up their throats and forcing air through them. If a frog can fill his throat with a lot of air, he can make a deep sound. If the frog is small and can fill his throat with only a little air, he can make a high-pitched sound. More air makes a lower sound. Less air makes a higher sound. Because of this, big frogs such as bull frogs make deep-sounding noises with their throats. Little frogs such as spring peepers and tree frogs make very high-pitched night noises. To imitate a frog, the children can use a toilet-paper cylinder and a paper-towel cylinder. Tell them to put a cylinder up to their mouth and try to sound like a frog. Make a recording of your classroom "night noises." As the children listen to it, ask them to close their eyes and pretend to be a little cricket hiding under a rock and listening or a tree frog listening while clinging to a tree. They might draw a picture of their classroom night noises.

☞ *Investigating vibrations*

Observing; designing investigations; applying knowledge

You can investigate air sounds by using six drinking glasses that are alike, water for filling the glasses, and a wooden dowel rod or pencil for striking the glasses. As you work, answer the questions below:

1. An empty glass is full of ______________ . (air)
2. Pour some water in one glass. The glass is now full of ______________ and ______________ . (air and water)
3. More air makes a lower sound. Pour a little bit of water in a glass. The glass is full of a little bit of water and a lot of ______________ . (air)
4. If you strike the glass gently with a wooden rod or pencil, what kind of sound does the glass make? (a lower sound)
5. Less air makes a higher sound. Pour a lot of water in the glass. The glass is full of a lot of water and a little bit of ______________ . (air)

6. Strike the glass gently with a wooden rod or pencil. What kind of sound does the glass make? (a higher sound)

Now fill the six glasses with different amounts of water. Before you strike each glass with a rod, can you predict whether it will produce a sound that is higher or lower? If you hear a very deep night noise, would you expect the animal that is making the noise to be big or little? (Big) If you hear a high-pitched night noise, would you expect the animal that is making the noise to be big or little? (Little)

Write or tell a story about going camping and listening to friendly night noises.

Let's create a paper-plate frog

Observing; following directions; creating a model; manipulating materials; measuring; finding information

Before creating this frog model (Fig. 4-23), observe a real frog, if possible, or examine some life-like frog models. Go to your media center and find books, posters, and videos about real frogs.

■ **4-23** *A paper-plate frog.*

What you need

- ☐ 9-inch white paper plate
- ☐ Scissors
- ☐ Crayons
- ☐ Markers
- ☐ Ruler
- ☐ White construction paper
- ☐ White glue

Directions

1. To create a frog model, first fold the paper plate in half.
2. Use the patterns in Fig. 4-24 to trace two back legs, two front legs, and two eyes onto white construction paper. Cut them out.
3. To make a long tongue for the frog, use a ruler to measure and draw a construction-paper strip $1/4$ by 8 inches. Curl the strip tightly around a pencil, and then remove the pencil. The strip will remain in a curled shape.

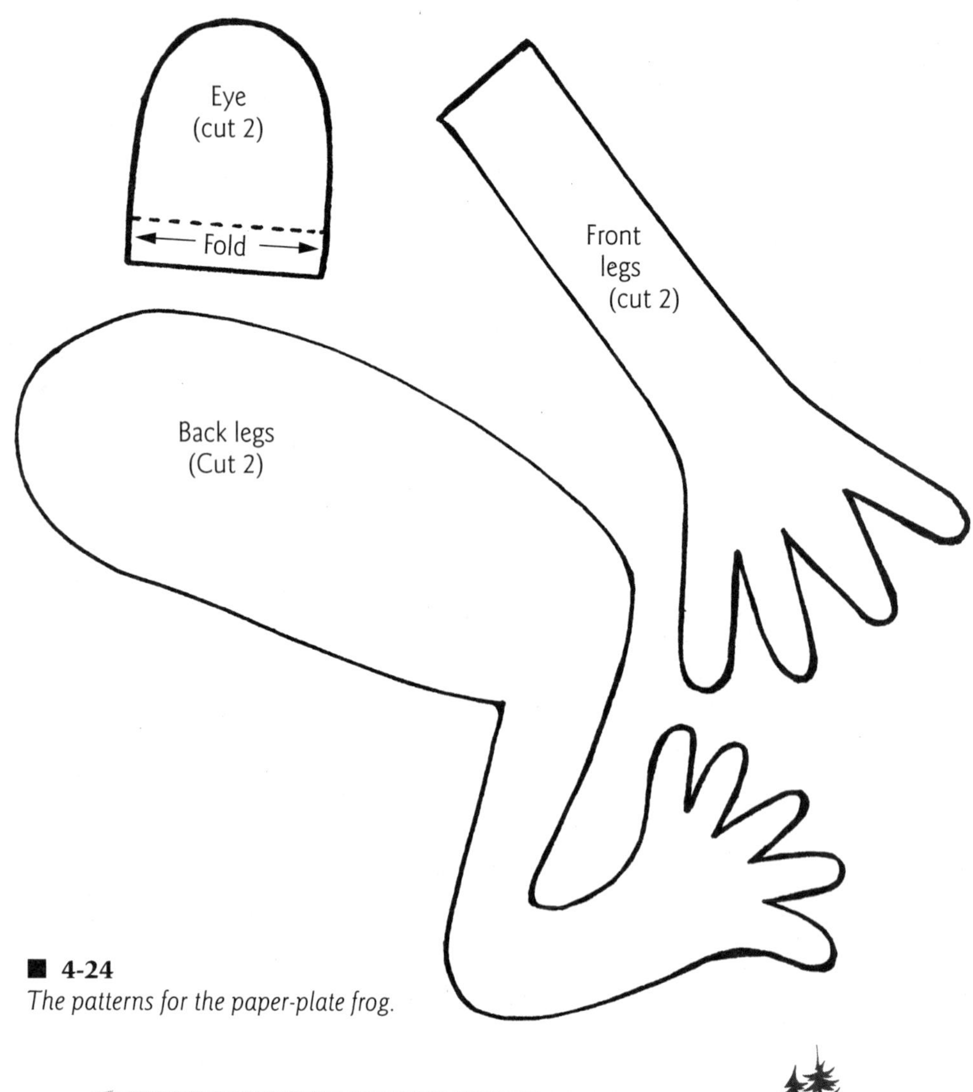

■ 4-24
The patterns for the paper-plate frog.

4. Look at a model or a picture of a real frog. Color your frog's body, legs, eyes, and tongue to resemble it. Draw big black eyeballs on one side of the eyes. Fold the bottom of each eye as shown on the pattern.
5. Glue the two back legs under the frog so that they stick out behind (see Fig. 4-23). Glue the two front legs under the frog so that they stick out in front.
6. Glue the two eyes on top of the frog's head so that they stick up. Glue the curled tongue inside the frog's mouth.

Let's create a foam tray flannel board

Observing; following directions; creating a model; manipulating materials; measuring; finding information

It's fun to have your very own flannel board! You can make up stories like *The Shy Little Cricket* to share with your friends

What you need

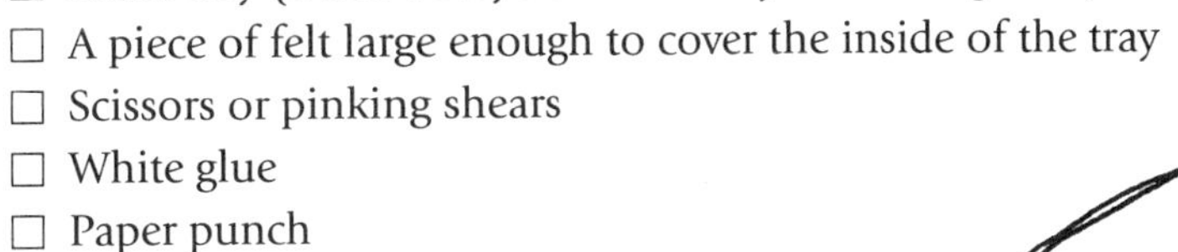

- ☐ Foam tray (a clean recycled meat tray from the grocery store)
- ☐ A piece of felt large enough to cover the inside of the tray
- ☐ Scissors or pinking shears
- ☐ White glue
- ☐ Paper punch
- ☐ Yarn
- ☐ Extra colored felt, including black and green

Directions

1. Cut the felt so that it fits inside the foam tray. Glue the felt in the tray (Fig. 4-25). Let the glue dry overnight.
2. Punch two holes in the top of the tray.

■ 4-25
A Styrofoam™ tray flannel board.

3. Measure a piece of yarn that is long enough to fit over your head. Thread the yarn through the holes and tie a knot.
4. Follow the patterns in Fig. 4-26 to trace and cut out a cricket from black felt, a leaf from green felt, and a footprint from felt of any color.

■ **4-26**
The patterns for the shy little cricket story.

5. Hang your foam tray flannel board over your head and tell the story of "The Shy Little Cricket." Make up new stories with your own felt characters.

"The Shy Little Cricket"

This poem can be used with the foam tray flannel board. Encourage children to make up their own flannel board stories.

The shy little cricket hops along
Softly singing a nighttime song.
Cool nights of spring and the chill of fall
Slow down the little cricket's call.
When warm summer nights come at last,
He sings his song very fast.
Now he hears a rustling sound;
Wait! What is it? What's coming around?
He stops singing so he can hear
The footsteps that are walking near.
Under a leaf he crawls to hide
Until the footsteps pass him by.
Then the shy little cricket creeps into sight
To sing his sweet song all through the night.

Frosty nights

During early spring and late fall, as temperatures drop at night, ice crystals often form on the ground. With the following activities, children can observe crystals as they create a frosty night camping scene and other crystal formations.

Let's create a frosty night camping scene

Using reference materials to find information about crystals; observing crystals; observing crystal growth; following directions; manipulating materials

Crystals are formed when conditions are right for molecules to line up in an orderly fashion. During the cold nights of early spring, ice crystals form on the ground, sometimes lifting loose dirt and sand onto glistening icy spikes that may be 2 inches long! You can create a frosty night camping picture by brushing your drawings with crystal water.

A note to parent or teacher: If you feel that heating water on an electric hot plate in the classroom is unsafe in your classroom setting, you can make the crystal solution at home and bring it to school in a closed jar. However, the more your students observe and take part in the making of the solution, the more they can learn. Epsom salt is a laxative. It should not be put in anyone's mouth. Children should wash their hands after handling *any* crystals.

What you need

- ☐ A book about crystals from your library or media center
- ☐ Magnifying lenses
- ☐ Salt
- ☐ Sugar
- ☐ Rock crystals purchased from a school supply or nature store (optional)
- ☐ Epsom salt (magnesium sulfate)
- ☐ Construction paper in dark colors of black, purple, or blue
- ☐ Crayons
- ☐ A hot plate
- ☐ A small pot for boiling water
- ☐ Measuring cups
- ☐ Wooden spoon
- ☐ A wide watercolor brush

Directions

1. Put three sheets of dark construction paper on a table. Spread a few crystals of salt on one paper, a few sugar crystals on another paper, and a few Epsom salt crystals on the third paper. Look at them with the magnifying lenses. How is each one different?
2. Make crystal water by mixing one cup Epsom salt and one cup water in a pot. An adult should heat the water to boiling and stir until the Epsom salt has dissolved. Let this mixture cool before using it.
3. Draw a nighttime camping scene with crayons on dark construction paper. Use light-colored crayons to make the picture visible. Press hard on the crayons to make the colors bright for the best results.
4. Brush cooled crystal water over your entire picture. Set your picture aside to dry. As the Epsom salt solution dries, it forms beautiful feathery crystals on the paper.
5. Examine these new crystals with a magnifier.

Extension activities

- ☐ Create a mystery crystal display, using salt and sugar on separate sheets of dark paper. See if you can tell which one is which by using only a magnifier.
- ☐ Crystals are very beautiful. Use reference books to find pictures of more crystals.
- ☐ Ask your parents to help you look for crystals around your home. Some people collect crystals for their own enjoyment. Many people wear crystals as jewelry.

Make a list of your discoveries.

Nighttime Notes *Create your own magnifiers by filling a cylindrical container with water and sealing it with a tight-fitting lid. Turn the container on its side and look through it. The water makes small things look larger. A clear film cartridge, a tall baby food jar, or a small mayonnaise jar make good magnifiers.*

Let's create bath salts

Observing crystals; observing crystal growth; following directions; manipulating materials

Bath salts are made of Epsom salt crystals. They make your bath water feel and smell good. Add a tablespoon to your bath water at home to soften it. The crystals in the plastic bag will grow larger as time passes because of the addition of the wet food color and cologne.

What you need

- ☐ A clean see-through jar with lid
- ☐ Epsom salt (1/4 cup per student)
- ☐ Liquid food color
- ☐ Bottle of inexpensive cologne
- ☐ Sandwich-size plastic bags or baby food jars

Directions

1. Help the children measure 1/4 cup Epsom salt into the jar. They can add a small squirt of food coloring and a little bit of cologne.
2. The child can put the lid on the jar and shake it gently to disperse the contents.
3. Pour the bath salts into the plastic bag. Create a label for the bath salts that says "Bath Salts: Add a tablespoon to your bath water to soften it."

Let's create fast crystals

Observing crystals; observing crystal growth; following directions; manipulating materials

Find out more about crystals by setting up your own crystal factory. Fast crystals are made of alum, which is used for making dill pickles.

A note to parent or teacher: Although alum is used when making pickles, it should not be put in anyone's mouth. Children should wash their hands after handling any crystals. This solution will crystallize in 45 minutes and therefore cannot be prepared in advance.

What you need

- ☐ 4 ounces powdered alum
  (available from a drug store or pharmacy)
- ☐ Dark-colored construction paper
- ☐ Magnifying lenses
- ☐ A recycled jar, such as a see-through peanut-butter jar
- ☐ 12 inches of string (or dental floss)
- ☐ Pencil
- ☐ 1 cup boiling water
- ☐ Liquid food color—green for making "emeralds" or red for "rubies"
- ☐ A plastic spoon for stirring

Directions

1. Spread a few crystals of alum on a sheet of dark construction paper. Look at them with a magnifier.
2. An adult should pour the alum in the jar in front of the children. A child can tie a 12-inch string onto a pencil so the string can be suspended in the jar.
3. An adult should pour 1 cup of boiling water in the jar. Stir to dissolve the alum. Stir in a large squirt of green or red food color.
4. Immediately lower the string into the hot solution. Push it down with the spoon. Balance the pencil across the top of the jar.
5. Crystals form while the solution cools. Do not move it for about 45 minutes.
6. Take the string out of the solution and lay it on a paper towel to dry. Look at the crystals with a magnifier.

Books of Interest About Nighttime and Nocturnal Animals

Informational books

Where Does the Sun Sleep?
Time Life for Children
Alexandria, Virginia

This book has questions and answers that relate to bedtime for children. It presents factual information in ways that young children will find meaningful.

Stargazers
by Gail Gibbons
Holiday House
New York, New York, 1992

This book shows children how they can be stargazers, who are people who watch the night sky and learn about stars.

Our Satellite: The Moon
Barron's Educational Series
Hauppauge, New York

This book provides detailed information about the moon. Younger children would enjoy the illustrations, but the text would probably be difficult.

The Sun: Our Very Own Star
by Jeanne Bendick
The Millbrook Press, Inc.
Brookfield, Connecticut, 1991

This book has beautiful illustrations and presents factual information in an understandable way for young learners.

What Makes a Shadow?
by Clyde Robert Bulla
Harper Collins
New York, New York, 1994

Children learn about shadows as they play and explore.

In the Middle of the Night
by Kathy Henderson
Macmillan
New York, 1992

This book helps children understand the many people, like nurses, astronomers, and trash collectors, who work throughout the night.

Night Creatures
by Susanne Santoro Whayne
Simon & Schuster
New York, New York, 1993

This excellent informational book has wonderful illustrations. Each animal is presented in its natural habitat, and meaningful information is provided about each nocturnal animal.

Out in the Night
by Karen Liptak
Harbinger House
Tucson, Arizona, 1989

This informational book helps children learn about the many nocturnal animals that share our earth, from the tropics of Southeast Asia to the meadows in the Alps.

The Book of North American Owls
by Helen Roney Sattler
Clarion Books
New York, New York, 1995

The text of this book is too difficult for young readers, but the watercolor illustrations are very appropriate.

A Time for Sleeping
by Ron Hirschi
Cobblehill Books
New York, New York, 1993

This book is filled with fantastic photographs of sleeping animals, like polar bears and sea otters.

Ways Animals Sleep
by Jane R. McCauley
National Geographic Society
Washington, DC, 1983

This book has marvelous photographs of how animals sleep. Readers see sea lions, koalas, raccoons, and many other animals.

Zoobooks: Owls
Zoobooks: Night Animals
Wildlife Education, Ltd.
San Diego, California, 1992

These paperback books are filled with interesting information. The books have full-page photographs of animals.

Story books

Owl Moon
by Jane Yolen
Philimel Books
New York, New York, 1987

This story is about a little girl and her father who go owling.

Owl Babies
by Martin Waddell
Candlewick Press
Cambridge, Massachusetts, 1992

Three baby owls wait for their mother to come home. They have fears much like children and greatly miss their mother. She does return, and the ending is very touching.

Stellaluna
by Janell Cannon
Scholastic
New York, New York, 1993

Stellaluna is a darling baby bat who is separated from her mother for a while and is taken care of by birds. She tries to be a bird until she joyfully discovers that she is a bat.

Follow the Moon
by Sarah Weeks
Harper Collins
Hong Kong, 1995

A careful young boy finds a baby sea turtle and lovingly guides it toward the ocean, rescuing it from danger.

Night Tree
by Eve Bunting
Harcourt Brace Jovanovich
New York, New York, 1991

The true spirit of Christmas is beautifully portrayed in this book.

Can't You Sleep, Little Bear?
by Martin Waddell
Candlewick Press
Cambridge, Massachusetts, 1988

Big Bear gently and lovingly helps Little Bear deal with his fears of the dark.

Peeping and Sleeping
by Fran Manushkin
Clarion Books
New York, New York, 1994

This bedtime story is nice for a young child. A little boy and his father go out on a spring night to see what is making a strange noise. They discover tiny frogs and have a warm, loving adventure together (ages 4-6).

Goodnight Moon
by Margaret Wise Brown
Harper Trophy
1947

This gently poetic story is very calming to children as they share in a little rabbit's goodnights to many things.

Mooncake
by Frank Asch
Prentice-Hall
New York, NewYork, 1983

Bear and Little Bird wish they could take a bite out of the moon.

Franklin in the Dark
by Paulette Bourgeois
Scholastic
New York, New York, 1986

Franklin is an adorable turtle who is afraid to go into his shell because he is afraid of the dark.

Going to Sleep on the Farm
by Wendy Cheyette Lewison
Dial Books
New York, New York, 1992

A young boy and his dad talk about how farm animals go to sleep. At the end of the story, the boy is fast asleep in his own bed.

Ira Sleeps Over
by Bernard Waber
Houghton Mifflin
Boston, Massachusetts, 1972

Ira gets invited over to Reggie's house to spend the night, and he has to work out the problem of needing his teddy bear.

Index

ABOUT THE AUTHORS

Rhonda Vansant, Ed.D. is an early childhood educator and consultant. Barbara L. Dondiego, M.Ed. is an author and educational consultant. Both live in Marietta, Georgia, and are the authors of four other books in the Science in Every Sense series: *Moths, Butterflies, Other Insects, and Spiders; Cats, Dogs, and Classroom Pets; Shells, Whales, and Fish Tails;* and *Seeds, Flowers, and Trees.*

ABOUT THE ILLUSTRATOR

Claire Kalish holds a bachelor of arts degree from Adelphi University. A former teacher, she is the owner of an Atlanta business, Table Fables.